VOLUME NINETY

CURRENT TOPICS IN MEMBRANES

Ion Transport and Membrane Interactions in Vascular Health and Disease

CURRENT TOPICS IN MEMBRANES, VOLUME 90

Series Editor

IRENA LEVITAN
Division of Pulmonary and Critical Care Medicine
University of Illinois at Chicago
Chicago, IL, United States

VOLUME NINETY

CURRENT TOPICS IN MEMBRANES

Ion Transport and Membrane Interactions in Vascular Health and Disease

Edited by

MICHAEL STUREK

Department of Anatomy, Cell Biology, & Physiology,
Indiana University School of Medicine,
Indianapolis, IN, United States

ELSEVIER

ACADEMIC PRESS
An imprint of Elsevier

Academic Press is an imprint of Elsevier
50 Hampshire Street, 5th Floor, Cambridge, MA 02139, United States
525 B Street, Suite 1650, San Diego, CA 92101, United States
The Boulevard, Langford Lane, Kidlington, Oxford OX5 1GB, United Kingdom
125 London Wall, London, EC2Y 5AS, United Kingdom

First edition 2022

Notices
Knowledge and best practice in this field are constantly changing. As new research and experience broaden our understanding, changes in research methods, professional practices, or medical treatment may become necessary.

Practitioners and researchers must always rely on their own experience and knowledge in evaluating and using any information, methods, compounds, or experiments described herein. In using such information or methods they should be mindful of their own safety and the safety of others, including parties for whom they have a professional responsibility.

To the fullest extent of the law, neither the Publisher nor the authors, contributors, or editors, assume any liability for any injury and/or damage to persons or property as a matter of products liability, negligence or otherwise, or from any use or operation of any methods, products, instructions, or ideas contained in the material herein.

ISBN: 978-0-323-99788-1
ISSN: 1063-5823

For information on all Academic Press publications
visit our website at https://www.elsevier.com/books-and-journals

Publisher: Zoe Kruze
Acquisitions Editor: Leticia Lima
Developmental Editor: Jhon Michael Peñano
Production Project Manager: Abdulla Sait
Cover Designer: Greg Harris

Typeset by STRAIVE, India

Contents

Contributors

Mouhamad Alloosh
Department of Anatomy, Cell Biology, & Physiology, Indiana University School of Medicine, Indianapolis, IN, United States

Olaf A. Bachkoenig
Gottfried Schatz Research Center, Molecular Biology and Biochemistry, Medical University of Graz, Graz, Austria

Jill Badin
ZOLL Medical Corporation, Chelmsford, MA, United States

Peter Bartels
Department of Pharmacology, University of California, Davis, CA, United States

Benjamin Gottschalk
Gottfried Schatz Research Center, Molecular Biology and Biochemistry, Medical University of Graz, Graz, Austria

Wolfgang F. Graier
Gottfried Schatz Research Center, Molecular Biology and Biochemistry, Medical University of Graz, Graz, Austria

Cristine L. Heaps
Department of Physiology and Pharmacology; Michael E. DeBakey Institute for Comparative Cardiovascular Science and Biomedical Devices, School of Veterinary Medicine and Biomedical Sciences, Texas A&M University, College Station, TX, United States

Junyoung Hong
Department of Pharmacology, University of California, Davis, CA, United States

Dandan Huang
Department of Pharmacology and Toxicology, School of Medicine, Virginia Commonwealth University, Richmond, VA, United States

Guangbi Li
Department of Pharmacology and Toxicology, School of Medicine, Virginia Commonwealth University, Richmond, VA, United States

Pengyang Li
Division of Cardiology, School of Medicine, Virginia Commonwealth University, Richmond, VA, United States

Pin-Lan Li
Department of Pharmacology and Toxicology, School of Medicine, Virginia Commonwealth University, Richmond, VA, United States

Roland Malli
Gottfried Schatz Research Center, Molecular Biology and Biochemistry, Medical University of Graz, Graz, Austria

Miguel Martín-Aragón Baudel
Department of Pharmacology, University of California, Davis, CA, United States

Mikaela L. McKenney-Drake
Butler University, Health Sciences Department, Pharmacy and Health Sciences, Indianapolis, IN, United States

Eric A. Mokelke
W.L. Gore & Associates, Inc., Flagstaff, AZ, United States

Manuel F. Navedo
Department of Pharmacology, University of California, Davis, CA, United States

Madeline Nieves-Cintrón
Department of Pharmacology, University of California, Davis, CA, United States

Eric A. Pereira da Silva
Department of Pharmacology, University of California, Davis, CA, United States

Stacey Rodenbeck
Department of Biology, Harding University, Searcy, AR, United States

Trevor S. Self
Department of Physiology and Pharmacology, School of Veterinary Medicine and Biomedical Sciences, Texas A&M University, College Station, TX, United States

Michael Sturek
Department of Anatomy, Cell Biology, & Physiology, Indiana University School of Medicine, Indianapolis, IN, United States

Sharanee P. Sytha
Department of Physiology and Pharmacology, School of Veterinary Medicine and Biomedical Sciences, Texas A&M University, College Station, TX, United States

Viktor Yarotskyy
Department of Pharmacology and Toxicology, School of Medicine, Virginia Commonwealth University, Richmond, VA, United States

Xinxu Yuan
Department of Pharmacology and Toxicology, School of Medicine, Virginia Commonwealth University, Richmond, VA, United States

Preface

Cell membranes with transmembrane ion channels, exchangers, and pumps tightly regulate the transport of ions across the membranes. A general physiology textbook shows the organelles sitting statically in the cells, almost giving the impression that they operate independently of each other, enabling flux into and out of the cytosol. It is now appreciated that the many functions and signaling by ion transport require fine orchestration of membrane interactions to maintain cellular health. The theme of this volume is vascular health and disease that may occur if ion transport and membrane interactions are dysfunctional. There is a range of topics, including new cellular and molecular methodology and imaging. We span enhanced health by exercise training and disease model approaches from *in vitro* to *in vivo* spontaneous disease to *in vivo* gene editing. There can be discrepant, contradictory approaches and points of view, but all approaches have strengths and weaknesses. All chapters present "food for thought" to guide future research on basic ion transport mechanisms and for therapeutic targets.

Introductory Chapter 1 provides an overview of the generic vascular tree to make the critical point that the typical diameter of the macrovascular conduit artery is >10-fold the diameter of the microvascular resistance artery (<200 μm diameter). The conduit artery is not just a "larger" artery. Although fundamental ion transport is similar, the conduit artery has key functional differences from the microvessel. Chapter 2 shows membrane interactions in cultured human umbilical vein endothelial cells, whereby Ca^{2+} influx preferentially feeds the mitochondria. Organelle-specific fluorescent Ca^{2+} sensors are powerful tools to test hypotheses rigorously. Lysosomal patch-clamp and super-resolution microscopy presented in Chapter 3 provide exciting data on lysosome fusion, which clearly state our theme of membrane interactions. Ca^{2+} release from lysosomes *via* TRPML1 channels in cultured coronary endothelial cells triggers the trafficking and fusion of lysosomes with multivesicular bodies to facilitate degradation. If lysosomal Ca^{2+} release is impaired, then the multivesicular bodies release exosomes from the endothelial cells that may contribute to coronary artery disease. Chapter 4 identifies a nanodomain of signaling molecules that regulates voltage-gated $Ca_V1.2$ channels in mesenteric artery smooth muscle cells. The nanocomplex can have a broad systemic impact because components are broadly expressed in other cells. Exciting future directions are to determine the effect of hyperglycemia on the complex and $Ca_V1.2$ channel activity. The search for vascular smooth muscle-specific regulation will enable design of therapeutics. Chapter 5 focuses on the progression of Ca^{2+} transporter dysfunction in coronary conduit vascular smooth muscles during stages of atherosclerosis in swine. There is an increase of Ca^{2+} transport *via* voltage-gated Ca^{2+} channels and the sarcoplasmic reticulum Ca^{2+} sequestration and release in the early stages of atherosclerosis, but there is a decrease of Ca^{2+} transport in the later stages of atherosclerosis. These findings debunk the dogma that vascular Ca^{2+} overload would occur uniformly in disease. The results from the swine model translated well to human cells from diseased, explanted hearts. Furthermore, the swine coronary muscle shows robust Ca^{2+} signaling *via* the endolysosome two-pore channel. The potential role of endolysosome Ca^{2+} signaling in extracellular vascular calcification is discussed.

An overview of K^+ channels in Chapter 6 provides the basis for interpretation of K^+ channel function in atherosclerosis and the effects of exercise training. This is a unique comparison of coronary macro- and microvascular smooth muscles, which shows a greater frequency of spontaneous transient outward currents (STOCs) of K^+ in healthy microvascular vs. conduit cells. In the microvessels of diabetic dyslipidemic swine with early atherosclerosis, Ca^{2+} sparks decreased in parallel with decreased STOCs in isolated cells. Chronic exercise restored Ca^{2+} sparks and STOCs. In Chapter 7, Kv channel function in microvascular smooth muscles in the coronary collateral-dependent myocardium distal to a conduit coronary artery occlusion is compared to cells from nonoccluded territory and conduit arteries. Overall, microvascular K^+ currents were greater than were conduit artery cell K^+ currents. Surprisingly, after exercise training, microvascular K^+ currents were decreased in cells from the collateral-dependent region compared with those in sedentary pigs. Implications for the seemingly contradictory K^+ channel regulation to optimize control of the basal arterial tone are discussed.

These chapters increase our understanding of ion transport and membrane interactions in vascular health and disease. It is hoped that these insights will guide future studies. Vascular disease remains a major cause of death worldwide, and ion transport mechanisms hold great promise as therapeutic targets.

Acknowledgments

I owe many thanks to the authors of this volume. The content of their chapters is excellent, and their responsiveness to editorial reviews was invigorating. We were a team working together to paint a clear picture of vascular membrane physiology and pathophysiology. I greatly appreciate the continuous encouragement from Dr. Irena Levitan of the University of Illinois at Chicago. The entire CTM editorial staff were especially patient with our progress and were phenomenally responsive to and timely with their reviews and production. Thank you so much.

Michael Sturek
Department of Anatomy, Cell Biology, & Physiology,
Indiana University School of Medicine,
Indianapolis, IN, United States

CHAPTER ONE

Introduction to ion transport and membrane interactions in vascular health and disease

Michael Sturek*
Department of Anatomy, Cell Biology, & Physiology, Indiana University School of Medicine, Indianapolis, IN, United States
*Corresponding author: e-mail address: msturek@iu.edu

Contents

Abstract

Cardiovascular disease is on the rise, partially due to the continued increase in metabolic syndrome. Advances in basic research on vascular ion transport have the potential to provide targets for therapeutic interventions. Vascular specificity, which includes different vascular beds having different characteristics and the macro- vs. microvasculature, is a vitally important variable in characterization of ion transport. At the cellular level, targeted fluorescent biosensors for Ca^{2+}, super-resolution microscopy, and organelle patch clamp electrophysiology enable more detailed studies. The "MetS/diabetes milieu" includes increased and decreased insulin, and increased glucose, increased LDL/HDL cholesterol and triglycerides, and increased blood pressure. The duration and severity of MetS/diabetes components certainly affect the vascular phenotype and ion transport and membrane interactions. A combination of in vivo animal models and in vitro cell models to study ion transport in MetS/diabetes conditions is optimal. Gene editing and selective pharmacological tools should be used after or in conjunction with characterization of ion transport in vascular health and disease phenotypes. This is critical to determining the causal role of Ca^{2+} signaling in modulation of vascular phenotype. The ion transport and membrane interactions that are measured are typically only a snapshot in time in these dynamic processes occurring over the progression of health and disease. It is imperative that this concept be considered in the planning of long-term studies of vascular disease, ion transport experiments, and interpretation of the data. Future directions for our contributors' research will advance the field.

Current Topics in Membranes, Volume 90
ISSN 1063-5823
https://doi.org/10.1016/bs.ctm.2022.09.006

The purposes of this brief introduction are to: (1) provide an overview of vascular function and study design to guide the characterization of ion transport in vascular health and disease; (2) highlight major contributions of the chapters within this framework; and (3) consider future directions.

1. Vascular specificity in health and disease

A major concept in this volume is vascular specificity, which includes different vascular beds having different characteristics and the macro- vs. microvasculature. For example, in response to low oxygen in the lung, pulmonary arterioles constrict to divert blood flow to better ventilated areas for exchange of oxygen. In contrast, the skeletal muscle vascular bed arterioles dilate in response to low oxygen to increase blood flow to working muscle. Several chapters in this volume contrast macro- vs. microvascular dynamics and *Current Topics in Membranes*, volume 85 focused on the microcirculation (Jackson, 2020). Fig. 1 shows the striking difference in pathology in the

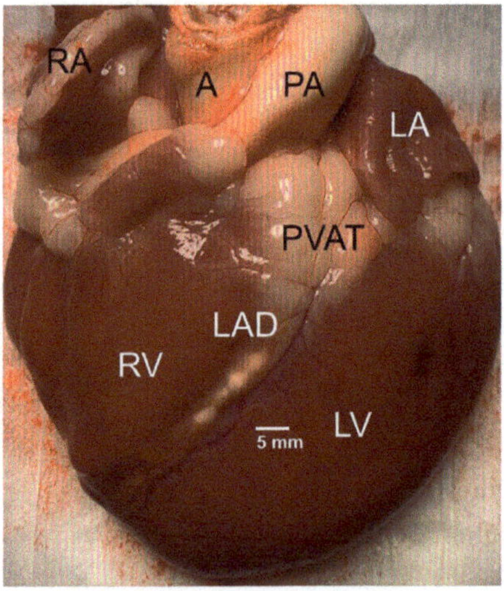

Fig. 1 Coronary circulation as an example of a vascular bed with macrovascular and microvascular components. The left anterior descending (LAD) coronary artery originates at the aorta (A), curves downward through the perivascular adipose tissue (PVAT), and angles left slightly on its way toward the bottom of the image. Immediately below the LAD label are four major atherosclerotic plaques. The LAD is a typical macrovascular conduit artery that is about 20-fold larger in diameter than the small veins and resistance arteries running diagonally and downward to the right of the LAD. Abbreviations: RA, right atrium; A, aorta; PA, pulmonary artery; LA, left atrium; PVAT, perivascular adipose tissue; LAD, left anterior descending artery; RV, right ventricle; LV, left ventricle.

coronary macro- vs. microvascular circulation. The left anterior descending (LAD) coronary artery is an epicardial conduit artery that is surrounded at the top of the heart by perivascular adipose tissue (PVAT). The artery is otherwise largely collapsed without perfusion pressure ex vivo, but four round masses of atherosclerotic plaque have almost occluded the artery below the LAD label. Diagonal branches of small veins and resistance arteries run across the left ventricle (LV). The LAD is a typical macrovascular conduit artery that is about 20-fold larger in diameter than the small veins and resistance arteries. See also the cross-sections of the schematic arteries in Fig. 2 where the microvessel to the right in the lumen of the conduit is only 5% the diameter of the conduit. A major difference from the conduit is that these microvessels have no detectable atherosclerosis. This difference cannot be explained by occlusion of the LAD artery that would have totally prevented blood flow, including atherogenic plasma cholesterol flow, to the microvessels. Such an occlusion would have created infarctions that should be seen as pale

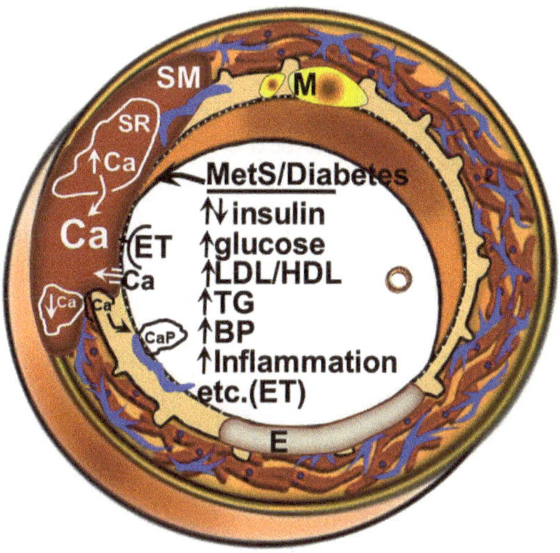

Fig. 2 Metabolic disease milieu bathes the vascular wall in conduit arteries and microvessels. All components of the metabolic disease milieu exist in the blood, but the significantly different concentrations are responsible for modulating vascular function. The smooth muscle (SM) and endothelial (E) cells are enlarged for emphasis and showing ion transport. In the remainder of the vascular wall blue structures are collagen fibrils and red structures are more appropriately sized smooth muscle cells to mimic a trichrome stained histological section. See text for more details. Symbols and abbreviations: ↑, increase; ↓, decrease; CaP, hydroxyapatite; MetS/D, metabolic syndrome/diabetes/disease; SM, smooth muscle; M, macrophage; SR, sarcoplasmic reticulum; ET, endothelin; LDL/HDL, low density lipoprotein/high density lipoprotein ratio; TG, triglycerides; BP, blood pressure; E, endothelium.

colored scarring on the left ventricle surface. The microvessels and conduit LAD artery are in the same vascular bed, but they are functionally different.

Several other examples of functional differences between microvessels and coronary conduit arteries explained by ion transport are: (1) Myogenic tone is much greater in microvessels compared to conduits (Muller, Myers, & Laughlin, 1993). (2) Voltage-gated Ca^{2+} channel expression is greater in microvascular coronary smooth muscle (CSM) compared to conduit CSM, as shown by patch clamp measures of current density (Bowles, Hu, Laughlin, & Sturek, 1997, 1998). (3) Ca^{2+} sparks produced from opening of ryanodine receptor channels in the sarcoplasmic reticulum (SR) Ca^{2+} store are frequent in microvascular CSM. The resulting spontaneous transient outward K^{+} current (STOC) frequency is ~17-fold greater in micro- vs. macrovascular CSM (Mokelke, Dietz, Eckman, Nelson, & Sturek, 2005) (see chapter "Specificity of Ca^{2+}-activated K^{+} channel modulation in atherosclerosis and aerobic exercise training" by Mokelke et al.). Clearly, there are guiding principles of vascular function, ion transport, and membrane interactions that are generalizable, but we must acknowledge the ample, compelling evidence of heterogeneity.

2. Metabolic disease milieu and experimental approaches and models

Cardiovascular disease is a leading cause of death worldwide. Significant improvements have been made in coronary heart disease by effective coronary interventions (e.g. stent angioplasty) and pharmacotherapy (e.g. plasma lipid lowering drugs). Unfortunately, improvement has plateaued, primarily due to increasing prevalence of metabolic syndrome (MetS), which leads to type 2 diabetes (Kleinert et al., 2018; Mokdad et al., 2001). Metabolic syndrome increases coronary artery disease severity 3-fold above those with no MetS (Kim et al., 2010) and increases the incidence of coronary heart disease mortality ~2.8-fold (Lakka et al., 2002). Fig. 2 shows the "MetS/Diabetes milieu" insulting the vascular wall and includes increased and decreased insulin, and increased glucose, dyslipidemia (increased LDL/HDL cholesterol, triglycerides [TG]), increased blood pressure (BP), inflammation, and other factors, e.g. endothelin (ET) (Hall, 2003). Smooth muscle (SM) and endothelial (E) cells are featured here, but several other cell types are important, including lymphocytes (Wolf & Ley, 2019), macrophages (M) (Tabas, Garcia-Cardena, & Owens, 2015), and perivascular adipose (McKenney et al., 2014; McKenney-Drake et al., 2017). The duration and severity of MetS/diabetes components certainly affect the vascular phenotype and ion transport and membrane interactions. Note that the insulin levels are increased in MetS as a compensatory mechanism to maintain euglycemia in the presence of target tissue insulin resistance. Eventually, with worsening insulin resistance, the insulin-producing pancreatic beta cells start to fail, plasma insulin levels decrease, and fasting and post-prandial glucose levels rise. The increase in fasting glucose to >126 mg/dL in humans marks the diagnosis of diabetes. See Sturek et al. for a graph of the temporal relationship of these variables (Sturek, Alloosh, & Sellke, 2020). Metabolic syndrome is commonly defined as the clustering of at least three of the components in the vascular compartment in Fig. 2 (Roberts, Hevener, & Barnard, 2013). These are the compelling reasons that this introduction largely focuses on MetS and diabetes as examples of disease states.

What are the appropriate models of MetS and diabetes for studies of ion transport? The most concise answer: any model that mimics the human syndrome optimally (Roberts et al., 2013; Sturek et al., 2020). It is not that simple. Several comprehensive reviews can guide decisions on the metabolic characteristics of animal models (Cluzel, Ryan, Herisson, & Caplice, 2022; Sturek et al., 2020; Varga, Harangi, Olsson, & Hansen, 2010). Other reviews are helpful for selection of the cardiovascular characteristics of the animal model and will depend on the vascular bed (Basu & Bornfeldt, 2020; Cluzel et al., 2022; Hamamdzic & Wilensky, 2013; Libby, 2015; Sorop et al., 2020; Sturek et al., 2020; Wang, Xu, Zhao, & Wang, 2020). Libby has argued that "... attempts to model a human disease in mice virtually guarantee failure" (Libby, 2015). This issue needs very serious consideration, since >95% of all biomedical research studies are conducted on mice (Roberts et al., 2009). If comparison of the coronary conduit and microvasculature in the same animal model is needed, then large animal models are essential (Cluzel et al., 2022; Sturek et al., 2020). In vivo spontaneous or chemically-induced disease and in vivo gene edited models have significant strengths (see chapters "Regulation of exosome release by lysosomal acid ceramidase in coronary arterial endothelial cells: Role of TRPML1 channel" by Li et al.; "Vascular $Ca_V1.2$ channels in diabetes" by Pereira da Silva et al.; "Multiphasic changes in smooth muscle Ca^{2+} transporters during the progression of coronary atherosclerosis" by Badin et al.; "Specificity of Ca^{2+}-activated K^+ channel modulation in atherosclerosis and aerobic exercise training" by Mokelke et al.; "K^+ channels in the coronary microvasculature of the ischemic heart" by Sytha et al.). These studies utilize acutely isolated, native cells from the animal or human. In vitro models are powerful for experimental control, as shown very well (see chapters "An unexpected effect of risperidone reveals a nonlinear relationship between cytosolic Ca^{2+} and mitochondrial Ca^{2+} uptake" by Bachkoenig et al.; "Regulation of exosome release by lysosomal acid ceramidase in coronary arterial endothelial cells: Role of TRPML1 channel" by Li et al.; "Vascular $Ca_V1.2$ channels in diabetes" by Pereira da Silva et al.; "Multiphasic changes in smooth muscle Ca^{2+} transporters during the progression of coronary atherosclerosis" by Badin et al.). Critical parameters for in vitro models are: (1) maintain the differentiated state of vascular cells in vitro as a healthy control and (2) use pathologically relevant concentrations of components of the Mets/diabetes milieu. For example, do not expose cells in vitro to a glucose concentration of 900 mg/dL, because this is incompatible with life and plasma glucose rarely reaches that level even transiently in poorly controlled diabetes (Chaturvedi, Stevens, Fuller, & Grp, 1995).

3. Snapshot of ion transport and membrane interactions

The ion transport and membrane interactions that are measured are typically only a snapshot in time in these dynamic processes occurring over the progression of health and disease. It is imperative that this concept be considered in the planning of long-term studies of vascular disease, ion transport experiments, and interpretation of the data. Fig. 3 shows a time course of vascular phenotypes during the progression of disease (Bender et al., 2009; Chobanian et al., 1982; Cluzel et al., 2022; Corbett et al., 1992; Edwards et al., 2010; McKenney-Drake et al., 2016; Sturek et al., 2020; Varga et al., 2010). At disease onset there is often an acute, transient decrease in vascular tone (blue solid line). This is also measured as a hyperemic phase in vivo, especially if type 1 diabetes has a rapid onset. The decrease in vascular tone is more reliable in animals that are made diabetic by chemical induction (Chobanian et al., 1982; Corbett et al., 1992). The more widely known phenotype is an increase in vascular tone, both macro- and microvascular, that occurs in early MetS and diabetes (Bender et al., 2009; Cluzel et al., 2022; Edwards et al., 2010; McKenney-Drake et al., 2016; Sturek et al., 2020; Varga et al., 2010). There is a continued increase in microvascular tone (blue dashed line), which can partially explain the hypertension phenotype at the whole animal level (red line). Chronic exercise or another

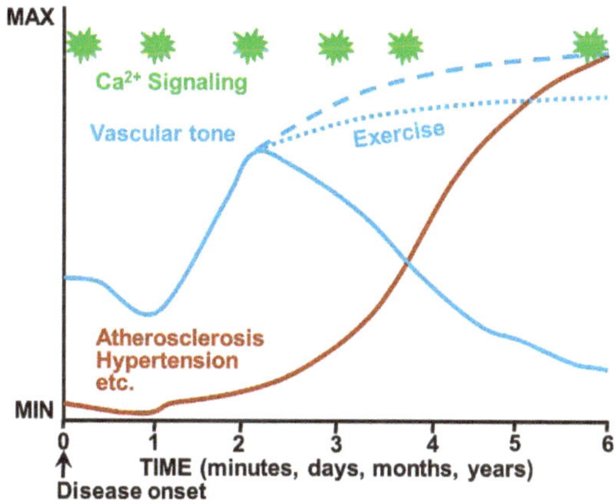

Fig. 3 Time course of Ca^{2+} signaling measures (ion transport and membrane interactions) and vascular phenotypes during the progression of vascular disease. The minimum (MIN) and maximum (MAX) are relative units for the measures. Time is relative and could have any of the units of time indicated. The Ca^{2+} signaling measures and events shown in green splatters occur at the times indicated. Vascular tone (blue lines) has three different outcomes: (1) The solid blue line that decreases most at time 6 represents the conduit artery. (2) The long dashed blue line shows the tone for diseased microvessels. (3) The dotted blue line shows the tone for diseased microvessels after chronic exercise training started at time 2.

therapeutic can be superimposed on the vascular disease state to attenuate the increase in vascular tone (blue dotted line) (Muller et al., 1993; Muller, Myers, & Laughlin, 1994).

The time course of changes in microvascular tone in Fig. 3 provides a basis for determining when Ca^{2+} signaling will be assessed. There is no continuous readout of Ca^{2+} signaling, so the green splatter marks in Fig. 3 are a rational starting point for Ca^{2+} signaling studies. Assume that healthy control animals (or in vitro cells) are matched for the duration, such that the healthy vascular phenotype would be a horizontal line in the vascular tone starting from the level at time 0. Shortly after disease onset at time 0 the assessment of Ca^{2+} signaling changes would represent the acute reaction to the disease milieu, i.e., MetS/diabetes milieu. The nadir in vascular tone at time 1 may represent the reaction to any of the initial Ca^{2+} signaling changes. In other words, the initial Ca^{2+} signaling changes could cause the nadir in vascular tone. This temporality of Ca^{2+} signaling preceding the vascular phenotype is highly consistent with Hill's principles of causation vs. association (Hill, 1965). We similarly showed that the increase in store-operated Ca^{2+} entry in CSM preceded atherosclerosis, which is consistent with a causal role of SOCE in CSM proliferation and migration in early stages (e.g., intimal thickening) of atherosclerosis (Dineen, Neeb, Obukhov, & Sturek, 2016). To frame our study within Fig. 3, we would have measured Ca^{2+} signaling changes at times 0 and/or 1 and atherosclerosis (red line) at times 2 or 3. The Ca^{2+} signaling changes at time 1 would definitely be associated with the nadir in vascular tone and potentially could cause the nadir. The Ca^{2+} signaling changes at time 2 would definitely be associated with the peak in vascular tone and potentially could cause the peak, but the Ca^{2+} signaling at time 2 certainly would not cause the nadir at time 1. Similar arguments can be made about temporality for the longer term changes from times 3–6 for microvascular tone and the effects of exercise. Chapter "Multiphasic changes in smooth muscle Ca^{2+} transporters during the progression of coronary atherosclerosis" by Badin et al. reviews the coronary conduit artery vascular tone and the progression of atherosclerosis, which are schematically shown in Fig. 3 by the blue solid line and the red line. The Ca^{2+} transporter function and dysfunction were characterized at times approximating times 0, 2, and 4 or 5. The Ca^{2+} transporter functional changes were biphasic, increasing in the early stage (time 2) and decreasing in the later stage (time 4 or 5) of atherosclerosis (McKenney-Drake et al., 2016). This is an example of detailed and definitive evidence of association, but additional tests of causation are needed.

The blunt question is: Why not just use gene editing to knock out or overexpress an ion transporter to show its role in vascular disease? Alternately, if you are desperate and do not have the gene edited cells or animals, why not use a "dirty" pharmacological antagonist or agonist? The compelling answer: characterizing the relatively more natural ion transporter function in vascular disease sets relevant boundaries for the ion transporter changes. For example, it would be somewhat premature for us to hypothesize that decreased endolysosome Ca^{2+} release via the TPC2 channel causes atherosclerosis and test the hypothesis by knocking out the channel. First, if endolysosome Ca^{2+} release via the TPC2 channel is not decreased in naturally occurring atherosclerosis, the knockout study has much less relevance. There is the exciting possibility of a robust vascular phenotype of the knockout, which would argue for the potential role of the TPC2 channel, but the result would not provide strong evidence for the "molecular mechanism of atherosclerosis". Second, even if TPC2 channel function is decreased in naturally occurring atherosclerosis, it is not likely to have zero activity as in the knockout state. One should consider the effect of heterozygous deletion as a more moderate impairment of overall TPC2 channel function. Gene overexpression is also more attractive to rescue TPC2 channel function if it was decreased in naturally occurring atherosclerosis. Taking these precautions is prudent for translating results of ion transport studies to therapies in the clinic. Having stated these precautions, it is absolutely clear that gene editing with CRISPR, etc. has revolutionized the field in all animal models, including large animals, which have been very challenging (Doudna, 2020; Tanihara, Hirata, & Otoi, 2021). Full genome sequencing has been accomplished for many large animal biomedical research models, such as Ossabaw miniature swine (Zhang et al., 2021), which have been widely used for ion transport research. Local delivery of genes for selective targeting in the vasculature in vivo has been achieved (Li et al., 2017). After characterizing the Ca^{2+} transporter functional changes, then target, target, target with gene editing to provide convincing evidence of causation.

4. Conclusions and future directions

The experimental tools to characterize ion transport and membrane interactions in health and disease are superb. It is fascinating to watch Ca^{2+} being transferred from the cytosol to organelles and the organelles delivering their molecular cargo elsewhere. Use of targeted fluorescent biosensors for Ca^{2+}, super-resolution microscopy, organelle patch clamp electrophysiology, proteomics, gene editing, etc. will certainly establish new paradigms and provide more detail about existing knowledge. The technical advances should be complemented by vital conceptual awareness of vascular specificity, the MetS/diabetes disease milieu, in vivo and in vitro models, and our ability to capture just snapshots of ion transport and membrane interactions during the progression of the vascular disease. Close attention to these concepts will work synergistically with technology to advance the field.

Acknowledgments

This work was supported by National Institutes of Health grant P30 DK097512. I thank all of my collaborators for their essential contributions to our work cited here.

References

Basu, D., & Bornfeldt, K. E. (2020). Hypertriglyceridemia and atherosclerosis: Using human research to guide mechanistic studies in animal models. *Frontiers in Endocrinology*, *11*, 504. https://doi.org/10.3389/fendo.2020.00504.

Bender, S. B., Tune, J. D., Borbouse, L., Long, X., Sturek, M., & Laughlin, M. H. (2009). Altered mechanism of adenosine-induced coronary arteriolar dilation in early-stage metabolic syndrome. *Experimental Biology and Medicine (Maywood, N.J.), 234*(6), 683–692. https://doi.org/10.3181/0812-RM-350.

Bowles, D. K., Hu, Q., Laughlin, M. H., & Sturek, M. (1997). Heterogeneity of L-type calcium current density in coronary smooth muscle. *American Journal of Physiology. Heart and Circulatory Physiology, 273*(42), H2083–H2089.

Bowles, D. K., Hu, Q., Laughlin, M. H., & Sturek, M. (1998). Exercise training increases L-type calcium current density in coronary smooth muscle. *American Journal of Physiology. Heart and Circulatory Physiology, 275*, H2159–H2169.

Chaturvedi, N., Stevens, L. K., Fuller, J. H., & Grp, T. W. M. S. (1995). Mortality and morbidity associated with body weight in people with IDDM: The WHO multinational study of vascular disease in diabetes. *Diabetes Care, 18*, 761–765.

Chobanian, A. V., Arquilla, E. R., Clarkson, T. B., Eder, H. A., Howard, C. F., Jr., Regan, T. J., et al. (1982). Cardiovascular complications. *Diabetes, 31*(suppl. 1), 54–64.

Cluzel, G. L., Ryan, P. M., Herisson, F. M., & Caplice, N. M. (2022). High-Fidelity porcine models of metabolic syndrome: A contemporary synthesis. *American Journal of Physiology. Endocrinology and Metabolism, 322*, E366–E381. https://doi.org/10.1152/ajpendo.00413.2021.

Corbett, J. A., Tilton, R. G., Chang, K., Hasan, K. S., Ido, Y., Wang, J. L., et al. (1992). Aminoguanidine, a novel inhibitor of nitric oxide formation, prevents diabetic vascular dysfunction. *Diabetes, 41*(4), 552–556. https://doi.org/10.2337/diab.41.4.552.

Dineen, S. L., Neeb, Z. P., Obukhov, A. G., & Sturek, M. (2016). Transient receptor potential channels in metabolic syndrome-induced coronary artery disease. In I. Levitan, & A. M. Dopico (Eds.), *Vascular ion channels in physiology and disease* (pp. 381–396). New York, NY: Springer (Reprinted from: In File).

Doudna, J. A. (2020). The promise and challenge of therapeutic genome editing. *Nature, 578*(7794), 229–236. https://doi.org/10.1038/s41586-020-1978-5.

Edwards, J. M., Neeb, Z. P., Alloosh, M. A., Long, X., Bratz, I. N., Peller, C. R., et al. (2010). Exercise training decreases store-operated Ca^{2+} entry associated with metabolic syndrome and coronary atherosclerosis. *Cardiovascular Research, 85*(3), 631–640. https://doi.org/10.1093/cvr/cvp308.

Hall, J. E. (2003). The kidney, hypertension, and obesity. *Hypertension, 41*(3 Pt 2), 625–633.

Hamamdzic, D., & Wilensky, R. L. (2013). Porcine models of accelerated coronary atherosclerosis: Role of diabetes mellitus and hypercholesterolemia. *Journal Diabetes Research, 2013*, 761415. https://doi.org/10.1155/2013/761415.

Hill, A. B. (1965). The environment and disease: Association or causation? *Proceedings of the Royal Society of Medicine, 58*, 295–300.

Jackson, W. F. (2020). Introduction to ion channels and calcium signaling in the microcirculation. *Current Topics in Membranes, 85*, 1–18. https://doi.org/10.1016/bs.ctm.2020.01.001.

Kim, J. Y., Mun, H. S., Lee, B. K., Yoon, S. B., Choi, E. Y., Min, P. K., et al. (2010). Impact of metabolic syndrome and its individual components on the presence and severity of angiographic coronary artery disease. *Yonsei Medical Journal, 51*(5), 676–682. https://doi.org/10.3349/ymj.2010.51.5.676.

Kleinert, M., Clemmensen, C., Hofmann, S. M., Moore, M. C., Renner, S., Woods, S. C., et al. (2018). Animal models of obesity and diabetes mellitus. *Nature Reviews. Endocrinology, 14*(3), 140–162. https://doi.org/10.1038/nrendo.2017.161.

Lakka, H. M., Laaksonen, D. E., Lakka, T. A., Niskanen, L. K., Kumpusalo, E., Tuomilehto, J., et al. (2002). The metabolic syndrome and total and cardiovascular disease mortality in middle-aged men. *JAMA, 288*(21), 2709–2716.

Li, W., Chen, X., Riley, A. M., Hiett, S. C., Temm, C. J., Beli, E., et al. (2017). Long-term spironolactone treatment reduces coronary TRPC expression, vasoconstriction, and atherosclerosis in metabolic syndrome pigs. *Basic Research in Cardiology, 112*. https://doi.org/10.1007/s00395-017-0643-0.

Libby, P. (2015). Murine "model" monotheism: An iconoclast at the altar of mouse. *Circulation Research, 117*(11), 921–925. https://doi.org/10.1161/CIRCRESAHA.115.307523.
McKenney, M. L., Schultz, K. A., Boyd, J. H., Byrd, J. P., Alloosh, M., Teague, S. D., et al. (2014). Epicardial adipose excision slows the progression of porcine coronary atherosclerosis. *Journal of Cardiothoracic Surgery, 9*, 2–12. https://doi.org/10.1186/1749-8090-9-2.
McKenney-Drake, M. L., Rodenbeck, S. D., Bruning, R. S., Kole, A., Yancey, K. W., Alloosh, M., et al. (2017). Epicardial adipose tissue removal potentiates outward remodeling and arrests coronary atherogenesis. *The Annals of Thoracic Surgery, 103*, 1622–1630. https://doi.org/10.1016/j.athoracsur.2016.11.034.
McKenney-Drake, M. L., Rodenbeck, S. D., Owen, M. K., Schultz, K. A., Alloosh, M., Tune, J. D., et al. (2016). Biphasic alterations in coronary smooth muscle Ca2 + regulation in a repeat cross-sectional study of coronary artery disease severity in metabolic syndrome. *Atherosclerosis, 249*, 1–9. https://doi.org/10.1016/j.atherosclerosis.2016.03.032.
Mokdad, A., Ford, E., Bowman, B., Nelson, D., Engelgau, M., Vinicor, F., et al. (2001). The continuing increase of diabetes in the U.S. *Diabetes Care, 24*, 412.
Mokelke, E. A., Dietz, N. J., Eckman, D. M., Nelson, M. T., & Sturek, M. (2005). Diabetic dyslipidemia and exercise affect coronary tone and differential regulation of conduit and microvessel K^+ current. *American Journal of Physiology. Heart and Circulatory Physiology, 288*(3), H1233–H1241. https://doi.org/10.1152/ajpheart.00732.2004.
Muller, J. M., Myers, P. R., & Laughlin, M. H. (1993). Exercise training alters myogenic responses in porcine coronary resistance arteries. *Journal of Applied Physiology, 75*, 2677–2682.
Muller, J. M., Myers, P. R., & Laughlin, M. H. (1994). Vasodilator responses of coronary resistance arteries of exercise-trained rigs. *Circulation, 89*, 2308–2314.
Roberts, C. K., Hevener, A. L., & Barnard, R. J. (2013). Metabolic syndrome and insulin resistance: Underlying causes and modification by exercise training. *Comprehensive Physiology, 3*, 1–58.
Roberts, R. M., Smith, G. W., Bazer, F. W., Cibelli, J., Seidel, G. E., Jr., Bauman, D. E., et al. (2009). Research priorities: Farm animal research in crisis. *Science, 324*(5926), 468–469.
Sorop, O., van de Wouw, J., Chandler, S., Ohanyan, V., Tune, J. D., Chilian, W. M., et al. (2020). Experimental animal models of coronary microvascular dysfunction. *Cardiovascular Research*. https://doi.org/10.1093/cvr/cvaa002.
Sturek, M., Alloosh, M., & Sellke, F. W. (2020). Swine disease models for optimal vascular engineering. *Annual Review of Biomedical Engineering, 22*, 25–49. https://doi.org/10.1146/annurev-bioeng-082919-053009.
Tabas, I., Garcia-Cardena, G., & Owens, G. K. (2015). Recent insights into the cellular biology of atherosclerosis. *Journal of Cell Biology, 209*(1), 13–22. https://doi.org/10.1083/jcb.201412052.
Tanihara, F., Hirata, M., & Otoi, T. (2021). Current status of the application of gene editing in pigs. *The Journal of Reproduction and Development, 67*(3), 177–187. https://doi.org/10.1262/jrd.2021-025.
Varga, O., Harangi, M., Olsson, I. A. S., & Hansen, A. K. (2010). Contribution of animal models to the understanding of the metabolic syndrome: A systematic overview. *Obesity Reviews, 11*, 792–807.
Wang, D., Xu, X., Zhao, M., & Wang, X. (2020). Accelerated miniature swine models of advanced atherosclerosis: A review based on morphology. *Cardiovascular Pathology, 49*, 107241. https://doi.org/10.1016/j.carpath.2020.107241.
Wolf, D., & Ley, K. (2019). Immunity and inflammation in atherosclerosis. *Circulation Research, 124*(2), 315–327. https://doi.org/10.1161/CIRCRESAHA.118.313591.
Zhang, Y., Fan, G., Liu, X., Skovgaard, K., Sturek, M., & Heegaard, P. M. H. (2021). The genome of the naturally evolved obesity-prone Ossabaw miniature pig. *iScience, 24*(9). https://doi.org/10.1016/j.isci.2021.103081.

CHAPTER TWO

An unexpected effect of risperidone reveals a nonlinear relationship between cytosolic Ca^{2+} and mitochondrial Ca^{2+} uptake

Olaf A. Bachkoenig, Benjamin Gottschalk, Roland Malli, and Wolfgang F. Graier*

Gottfried Schatz Research Center, Molecular Biology and Biochemistry, Medical University of Graz, Graz, Austria

*Corresponding author: e-mail address: wolfgang.graier@medunigraz.at

Contents

Current Topics in Membranes, Volume 90
ISSN 1063-5823
https://doi.org/10.1016/bs.ctm.2022.09.001

Abstract

Mitochondria actively contribute to cellular Ca^{2+} homeostasis. The molecular mechanisms of mitochondrial Ca^{2+} uptake and release are well characterized and are attributed to the multi-protein assembly of the mitochondrial Ca^{2+} uniporter complex (MCUC) and the mitochondrial sodium-calcium exchanger (NCLX), respectively. Hence, Ca^{2+} transfer from the endoplasmic reticulum (ER) and store-operated Ca^{2+} entry (SOCE) into the mitochondrial matrix has been quantitatively visualized on the subcellular level using targeted fluorescent biosensors. However, a correlation between the amplitude of cytosolic Ca^{2+} elevation with that in the mitochondrial matrix has not been investigated in detail so far. In the present study, we combined the Ca^{2+}-mobilizing agonist histamine with the H_1-receptor antagonist risperidone to establish a well-tunable experimental approach allowing the correlation between low, slow, high, and fast cytosolic and mitochondrial Ca^{2+} signals in response to inositol 1,4,5-trisphosphate (IP_3)-triggered ER Ca^{2+} release. Our present data confirm a defined threshold in cytosolic Ca^{2+}, which is necessary for the activation of mitochondrial Ca^{2+} uptake. Moreover, our data support the hypothesis of different modes of mitochondrial Ca^{2+} uptake depending on the source of the ion (i.e., ER vs SOCE).

Data availability

The data that support the findings of this study are available from the authors on reasonable request.

1. Introduction

For a long time, mitochondria were thought to serve as passive Ca^{2+} sinks that store Ca^{2+} by accumulating Ca^{2+} polyphosphate in the matrix. However, with the introduction of mitochondria-targeted Ca^{2+} biosensors (Rizzuto, Simpson, Brini, & Pozzan, 1992) and the pioneering work of Tullio Pozzan and Rosario Rizutto (Pozzan & Rizzuto, 2000; Rizzuto et al., 1992) this view has dramatically changed. Today, mitochondria are well known to be actively involved in cellular Ca^{2+} homeostasis in health and disease (Modesti et al., 2021). However, the topic of mitochondrial Ca^{2+} homeostasis slowly developed until the landmarking work on the identification of the mitochondrial calcium uptake 1 (MICU1) by Fabiana Perrocchi in Vamsi K. Mootha's group who described MICU1 as part of the mystic mitochondrial Ca^{2+} uniporter (Perocchi et al., 2010). The discovery of MICU1 paved the way for many subsequent findings, like the mitochondrial Ca^{2+} uniporter protein MCU (Baughman et al., 2011; Stefani, Raffaello, Teardo, Szabò, & Rizzuto, 2011), the essential MCU regulator (EMRE) (Sancak et al., 2013), and the mitochondrial calcium uniporter regulator 1 (MCUR1) (Mallilankaraman et al., 2012) to name the most prominent. Consequently, mitochondrial Ca^{2+} uptake is now known to be achieved by a highly controlled multiprotein complex, the mitochondrial Ca^{2+} uniporter complex (MCUC) (Feno, Rizzuto, Raffaello, & Reane, 2020; Mammucari, Gherardi, & Rizzuto, 2017; Marchi & Pinton, 2014). MICU1 and its paralogs MICU2 (Ahuja & Muallem, 2014; Foskett & Madesh, 2014; Kamer & Mootha, 2014; Plovanich et al., 2013) and, in primary cortical neurons (Patron, Granatiero, Espino, Rizzuto, & Stefani, 2018) and cardiomyocytes (Puente et al., 2020) MICU3, have been found as the most important regulator of MCUC (Csordás et al., 2013; Foskett & Madesh, 2014). Under low/free Ca^{2+} conditions MICU1 prevents MCUC activity by the formation of hexamers and undergoes conformational changes to non-inhibiting oligomers in a higher Ca^{2+}-containing environment (Wang et al., 2014). By measuring förster-resonance energy transfer (FRET) within CFP and YFP labeled MICU1, our team measured an EC_{50} for Ca^{2+}-induced rearrangement of MICU1 of 4.4 μM (Waldeck-Weiermair et al., 2015). While Ca^{2+} levels in the global cytosol seldomly exceed 1 μM independently of the origin of Ca^{2+}, at the surface of mitochondria, Ca^{2+} hotspots with >14 μM develop upon intracellular Ca^{2+} release (Giacomello et al., 2010), thus, allowing Ca^{2+}-induced rearrangement of the MICU1 multimers, and, subsequently, the activation of the MCUC (Waldeck-Weiermair et al., 2015; Wang et al., 2014). Applying simultaneous measurable (sub-organellar) targeted genetically encoded Ca^{2+} sensors (Waldeck-Weiermair et al., 2012, 2019), the individual Ca^{2+} kinetics of the mitochondrial sub-compartments upon the Ca^{2+} transfer from the ER toward the mitochondrial matrix were revealed and point to a complex integration of multiple processes in this phenomenon (Gottschalk et al., 2019). Notably, the kinetics and possibly the underlying molecular machinery of mitochondrial Ca^{2+} uptake critically depends on the source of Ca^{2+} approaching the organelle's surface (Waldeck-Weiermair et al., 2011; Waldeck-Weiermair, Malli, et al., 2010). However, despite these insights in mitochondrial Ca^{2+} homeostasis and the organelle's contribution to cellular Ca^{2+} homeostasis, a correlation between the elevation of cytosolic Ca^{2+} concentrations with that of the mitochondria has not been investigated in detail so far.

In the present study, we took benefit of the potency of the atypical antipsychotic risperidone to inhibit H_1 receptors (Leysen et al., 1988, 1992) that are the mediators of histamine-triggered Ca^{2+} signaling in endothelial cells (Spitaler, Hammer, Malli, & Graier, 2002; Voorde & Leusen, 1983). We found that risperidone concentration-dependently attenuates histamine but not ATP-induced Ca^{2+} signals in the endothelial cell line EA.hy926 (Edgell, McDonald, & Graham, 1983), and utilized this effect to explore the interdependency and correlation of agonist-induced cytosolic Ca^{2+} elevations with that in the mitochondrial matrix. Our data support the concept of a rather high threshold for the Ca^{2+} transfer into the mitochondria and point to a nonlinear correlation between cytosolic and mitochondrial Ca^{2+} signals. Moreover, risperidone was found to act as an antagonist on the endothelial H1 receptor at concentrations potentially considerable during antipsychotic human therapy.

2. Materials and methods

2.1 Materials and compounds

Desfluoro Risperidone (Cat. D290065) was bought from Toronto Research Chemicals (Toronto, Canada), dissolved in DMSO, aliquoted and stored at −20 °C. Buffer salts, ATP (Art.No. HN35.2), and histamine dihydrochloride (Art.No. 4017.6) were from Carl Roth (Karlsruhe, Germany). Fura-2/am (Cat. ab120873) was obtained from Abcam (Cambridge, UK). The Transfection Reagent (Polyjet™; SL100688) was purchased at SigmaGen® Laboratories (Fredrick, MD, USA). Cell culture plasticware was from Greiner Bio-One (Reinbach im Mühlkreis, Austria). Fetal calf serum (FCS), medium supplements, and DMEM were from ThermoFisher Scientific (Karlsruhe, Germany) and Sigma Aldrich (Vienna, Austria), respectively.

2.2 Cell culture and preparation

In this work, the endothelial cell line, EA.hy926 (Edgell et al., 1983), was used. Cells were cultured in DMEM with 10% FCS, 100 U/mL penicillin, 100 µg/mL streptomycin, and 1.25 µg/mL amphotericin B at 37 °C/5% CO_2. For imaging purposes, cells were seeded on 30 mm glass coverslips (VWR, Vienna, Austria) that were mounted in a perfusion chamber (PC30; NGFI, Graz, Austria; https://www.ngfi.eu) for microscopic measurements.

2.3 Transfection

Cells were seeded in six well plates 48h prior to transfection with the transfection reagent to a confluence of 50% under standard conditions as previously described (Oflaz et al., 2021). Prior to transfection, the medium was changed to DMEM without FCS and antibiotics. 5h after transfection medium was changed back to the full culture medium.

2.4 Buffers and solutions

For the storage prior to measurements and fura-2 loading a buffer (loading buffer) was used containing in mM: 2 $CaCl_2$, 138 NaCl, 5 KCl, 1 $MgCl_2$, 1 HEPES, 2.6 $NaHCO_3$, 0.44 KH_2PO_4, 0.34 Na_2HPO_4, 10 D-glucose, 0.1% vitamins, 0.2% essential amino acids and 1% penicillin/streptomycin. During experiments, cells were perfused with buffers containing physiological ion concentrations as follows in mM: 2 $CaCl_2$, 138 NaCl, 5 KCl, 1 $MgCl_2$, 1 HEPES, 10 D-glucose. In the case of 0 Ca buffer, no $CaCl_2$ but 0.1mM EGTA was added. pH was adjusted to 7.4 using NaOH.

2.5 Loading with fura-2/am

For cytosolic measurements, cells were loaded with fura2/am (Graier, Kurt, & Kukovetz, 1990). Cells were seeded 24–48h prior to experiments in 6-well plates on 30mm glass coverslips and cultured under standard conditions. Loading buffer was used to prepare a 1.66μM fura-2 solution and cells were incubated in 2mL for 20min at room temperature. After the incubation cells were washed two times with the loading buffer and stored at room temperature protected from light until the measurements (Graier, Simecek, & Sturek, 1995).

2.6 Genetically-encoded fluorescent sensors

To investigate subcellular Ca^{2+} levels different genetically encoded FRET sensors were used. These sensors were encoded in a eukaryotic expression vector, which was introduced into the cells via transfection. 4mtD3cpv (Ex: 430, Em: 480/535nm) was used for mitochondrial matrix measurements, IMS-GemGeco for measurements close to the IBM, and D1ER (Ex: 430, Em: 480/550nm) in the ER (Koshenov et al., 2022).

2.7 Microscopes and acquisition

For measurements, two microscopic devices were used (Koshenov et al., 2022). The Anglerfish (NGFI, Graz, Austria; www.ngfi.eu) equipped with a Nikon CFI Super Fluor 40× oil immersion NA1.3 (Nikon, Vienna, Austria), 2 CMOS cameras (acA1920-40um, Basler, Ahrensburg, Germany), 3 LEDs (345 nm/55 mW; 380 nm/>1500 mW; 470 nm/>900 mW; NGFI), and the respective set of filters (340/26, 380/14, and 480/17 nm; Semrock from AHF, Tuebingen, Germany) and dichroics (F38-366 and F38-495 dichroics; Semrock) was used. Moreover, a conventional inverted microscope (Olympus IX73, Olympus, Vienna, Austria) equipped with an ApoN340 40× oil immersion objective (Olympus), CCD QImaging Retiga R1 camera (Teledyne Photometrics, Tucson, AZ, USA), a LedHUB (Omnicorn Laserage Laserprodukte, Rodgau-Dudenhofen, Germany) with a 385 and 455 LED in combination with a CFP/YFP/RFP filter set (Olympus). In case of FRET measurements, the emission was split onto two halves of the camera using a dual-channel beam splitter DV2 (Teledyne Photometrics) and a CFP/YFP beamsplitter (505DCXR, Chroma Technologies, Bellows Falls, VT, USA). For fura-2 measurements, cells were sequentially illuminated with 340 nm and 380 nm and emission was recorded at 510 nm. Data acquisition was done using VisiView 4.2.01 (Visitron Systems, Puchheim, Germany) or AnglerFish Acquisition Software 4/22 (NGFI) at the Olympus and AnglerFish, respectively. Both microscopes were coupled with a gravity-based perfusion system (PS-9D, NGFI).

2.8 Data analysis

The measurement was background-subtracted using a background ROI and corrected for bleaching using an exponential decay fit. ImageJ (NIH, Bethesda, MA, USA) and Microsoft Excel (Microsoft) were used for the image analysis, data extraction, and calculations (Malli, Frieden, Hunkova, Trenker, & Graier, 2007).

2.9 Statistics

Data shown represent the mean of a minimum of three experiments in triplicate on at least three different days (n) ± SEM. All experiments are repeated on at least three different days and numbers indicate the numbers of wells/single cells. Analysis of variance (ANOVA) with Tukey post hoc test was performed. GraphPad Prism software version 9.3.1 (GraphPad Software, San Diego, CA, USA) was used for the analysis and presentation of the data.

3. Results

Our previous approaches provided strong data in regard of the impact of the Ca^{2+} source and on-kinetics of Ca^{2+} diffusing toward the mitochondrial surface, but failed to provide a correlation between the cytosolic Ca^{2+} elevations with mitochondrial Ca^{2+} sequestration (Waldeck-Weiermair, Duan, et al., 2010; Waldeck-Weiermair et al., 2011; Waldeck-Weiermair, Malli, et al., 2010). Accordingly, we aimed to establish an experimental approach that allows us to correlate initial cytosolic Ca^{2+} elevations and mitochondrial Ca^{2+} sequestration due to IP_3-triggered ER Ca^{2+} release under well-defined tunable conditions.

3.1 Risperidone inhibited cytosolic Ca^{2+} transients

As expected, risperidone inhibited the histamine-induced cytosolic Ca^{2+} signaling in the human umbilical vein endothelial cell line EA.hy926 in a concentration-dependent manner (Fig. 1A). Risperidone reduced all phases of the histamine-induced Ca^{2+} signal, the initial peak, and the subsequent plateau phase. Therefore, the inhibitory potency of risperidone was calculated using the area under the curve (AUC) of the histamine-induced cytosolic Ca^{2+} signal (Fig. 1B). Our experiments revealed an IC_{50} for risperidone to prevent histamine-induced cytosolic Ca^{2+} elevation in endothelial cells of 35.74 (27.42–46.58) μM (Fig. 1C). The Hill coefficient for the respective curve fit was found to be 1.01. We investigated if risperidone potentially inhibits ER Ca^{2+} refilling. Therefore, the cytosolic Ca^{2+} elevations in response to two sequential histamine stimulations with interim Ca^{2+} refilling were measured. Upon the second histamine stimulation, a slight decrease in cytosolic Ca^{2+} peak response was observed, while the plateau phases following the peaks were comparable (Fig. 2A and B). Next, we investigated if the order of risperidone treatment would influence the effect of risperidone on histamine-induced ER Ca^{2+} release. Accordingly, risperidone was added during the first or second histamine stimulation. These experiments revealed that risperidone, independently a previous stimulation, reduced the cytosolic Ca^{2+} signal in response to histamine (Fig. 2C and D).

3.2 The effect of risperidone on endothelial Ca^{2+} was exclusively due to its inhibitory effect on histamine-triggered signaling

Next, we explored the underlying reason for the inhibitory effect of risperidone on endothelial Ca^{2+} signaling in more detail. We tested whether the effect of risperidone is limited to histamine or can also be found for another IP_3-generating agonist. Therefore, the effect of risperidone on intracellular Ca^{2+} release and subsequent store-operated Ca^{2+} entry (SOCE) in response to histamine and ATP was studied in series in one given cell. As illustrated in

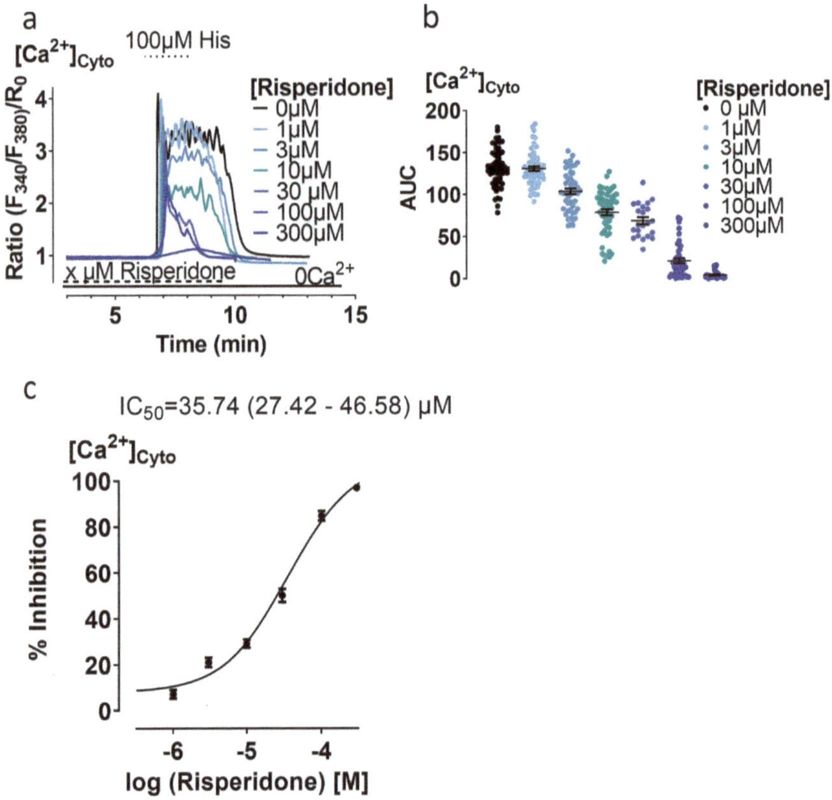

Fig. 1 Risperidone inhibits histamine-triggered cytosolic Ca^{2+} signals in fura-2 loaded single endothelial cells in a concentration-dependent manner. (A) Representative traces of the cytosolic Ca^{2+} signaling in single endothelial cells in response to 100 μM histamine in the absence and presence of 1–300 μM risperidone. (B) Summary of the effect of various risperidone concentrations on histamine-induced cytosolic Ca^{2+} signaling in single endothelial cells analyzed as the area under the curve (AUC) (concentration in μM risperidone-wells/cells: 0–16/58; 1–9/36; 3–15/36; 10–12/33; 30–10/22; 100–11/45; 300–6/35). (C) Concentration-response curve of risperidone on cytosolic Ca^{2+} elevations (AUC/panel b) in single endothelial cells.

Fig. 3A, cells were initially stimulated by 100 μM histamine in the nominal absence of extracellular Ca^{2+}. After a short wash-out we reintroduced extracellular Ca^{2+} to refill the endoplasmic reticulum (ER) via the SOCE. Subsequently, cells were stimulated with 100 μM ATP in the absence of extracellular Ca^{2+} followed by the re-addition of extracellular Ca^{2+} in the absence of ATP. Risperidone was always present 5 min prior to agonist addition until its washout. These experiments revealed a selective inhibition of histamine-induced Ca^{2+} signal via the H_1 receptor (Spitaler et al., 2002)

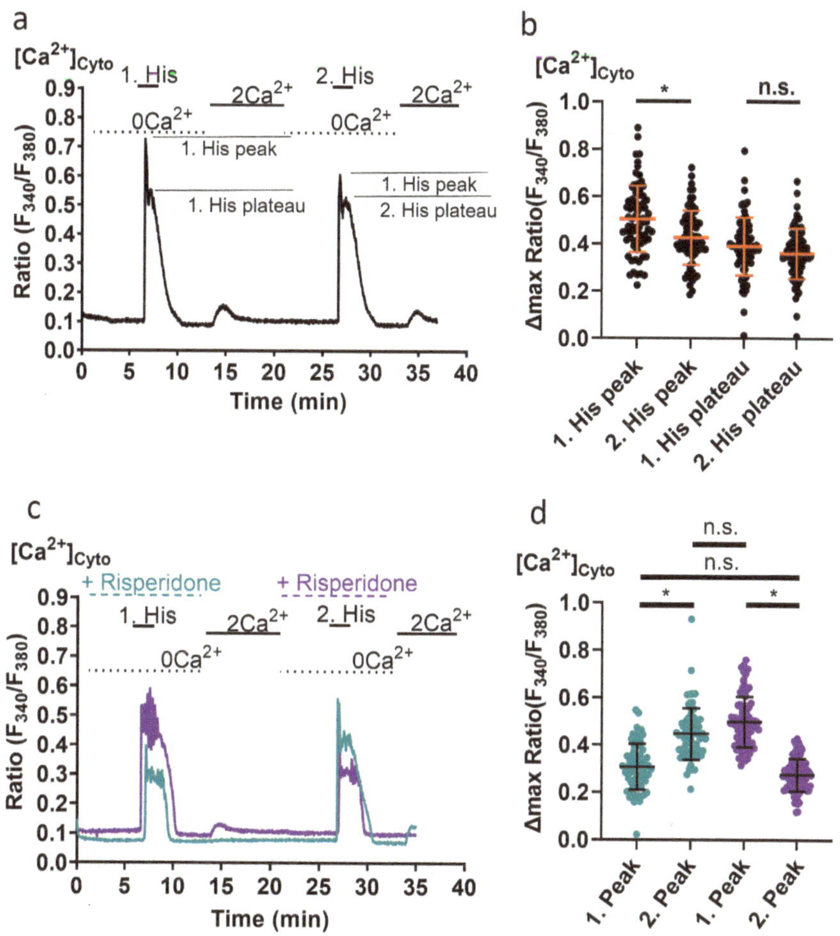

Fig. 2 Risperidone does not imped ER Ca^{2+} refilling. (A) Representative traces of the cytosolic Ca^{2+} signaling in single endothelial cells in response to two sequential stimulations with 100 μM histamine. (B) Statistical evaluation of Δmax peak and plateau (indicated in A) measurements of two sequential histamine stimulations. Error bars indicate mean +/− SD (1.His stimulation: wells/cells 4/76; 2.His stimulation: 4/73). (C) Representative traces of the cytosolic Ca^{2+} signaling in single endothelial cells in response to two subsequent stimulations with 100 μM histamine in the absence and presence of 10 μM risperidone. As indicated, cells were stimulated by the agonist in the presents of risperidone during the first or second histamine stimulation. (D) Statistical evaluation of Δmax peak measurements of two subsequent histamine stimulations with risperidone present during the first (turquoise blue) or second histamine stimulation. Error bars indicate mean +/− SD (risperidone 1.His: wells/cells 4/71; risperidone 2.His: 5/91).

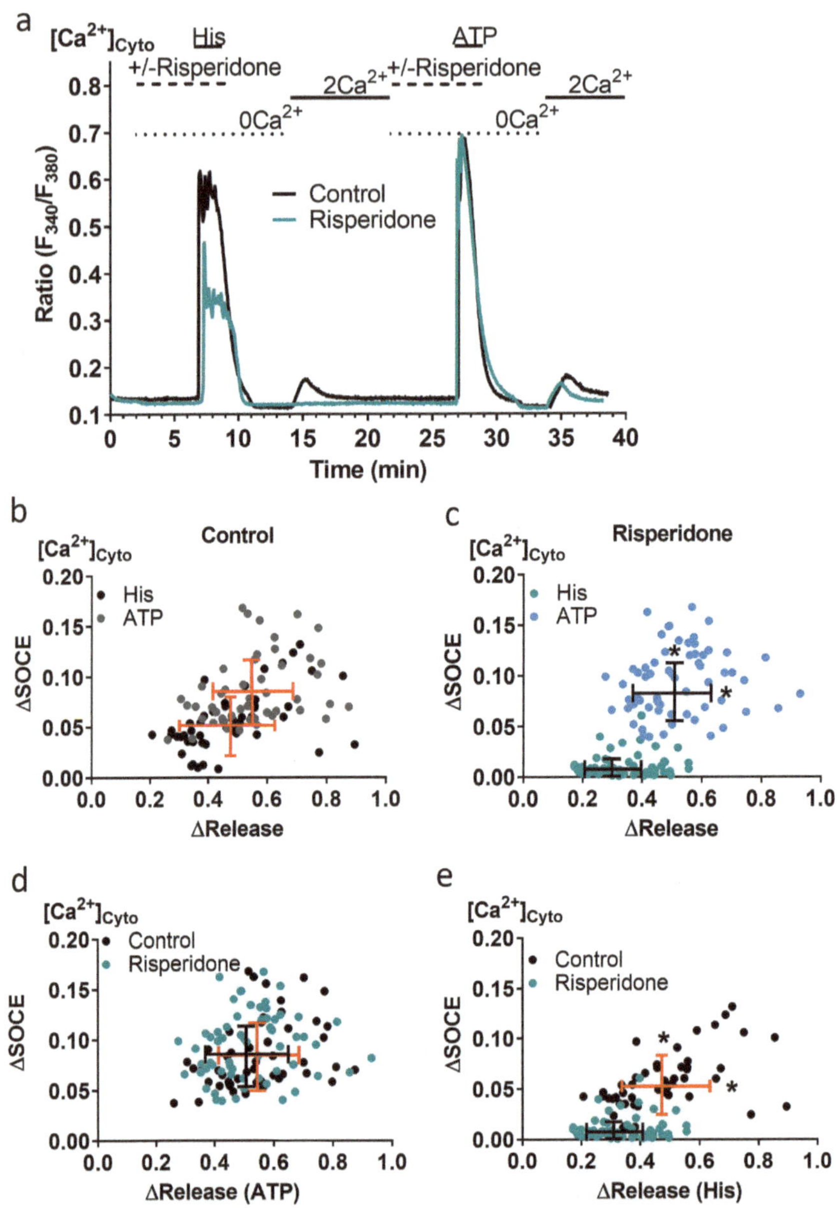

Fig. 3 Comparison of the effect of risperidone on cytosolic Ca^{2+} signals in single endothelial cells serially stimulated with histamine and ATP. (A) Representative traces of the cytosolic Ca^{2+} signaling in single endothelial cells in response to 100 μM histamine and subsequently 100 μM ATP in the absence and presence of 10 μM risperidone. As indicated, cells were stimulated by the agonist in the absence of extracellular Ca^{2+} (Release) followed by the addition of extracellular Ca^{2+} after agonist washout (SOCE). (B–E) The changes in the Ratio were calculated as maximum delta (Δ). Correlations between the agonist-induced Ca^{2+} elevation in the absence of extracellular Ca^{2+} (ΔRelease) and the SOCE after readdition of extracellular Ca^{2+} after agonist washout (ΔSOCE). Error bars indicate mean +/− SD (control: wells/cells 9/49; risperidone: 9/63).

in the endothelial cells (Fig. 3A). The correlation between the intracellular Ca^{2+} release (ΔRelease) with the subsequent Ca^{2+} increase due to the SOCE (ΔSOCE) provided similar results for histamine and ATP (Fig. 3B). However, in the presence of 10 μM risperidone, histamine-induced intracellular Ca^{2+} release and subsequent SOCE were abolished (Fig. 3C and D), while ATP-triggered Ca^{2+} signals remained unaffected (Fig. 3C and E).

3.3 Risperidone inhibited histamine-induced mitochondrial Ca^{2+} signals

Next, we investigated how risperidone affects histamine- and ATP-induced Ca^{2+} signals in the mitochondria. Therefore, endothelial cells were transiently transfected with the mitochondrial-targeted, genetically encoded Ca^{2+} sensor 4mtD3cpv (Palmer et al., 2006; Palmer & Tsien, 2006), and the identical protocol mentioned above (i.e., Fig. 2A) was performed. Like in the cytosol, risperidone prevented histamine-induced Ca^{2+} elevation in the mitochondrial matrix, while ATP-induced mitochondrial Ca^{2+} elevation was not affected (Fig. 4A). The statistical analysis revealed a strong effect of risperidone on histamine-induced initial mitochondrial Ca^{2+} elevation and subsequent SOCE-dependent mitochondrial Ca^{2+} signal but no effect on the respective signals upon cell stimulation with ATP (Fig. 4B and C). Notably, when using 1 μM histamine it became obvious that the inhibitory effect of risperidone on mitochondrial Ca^{2+} elevations appeared stronger than that on the respective cytosol Ca^{2+} elevation (Fig. 4D). Therefore, we obtained a concentration-response curve for risperidone on histamine-induced Ca^{2+} signals in the mitochondrial matrix (Fig. 4E) and extracted the respective maximal mitochondrial Ca^{2+} elevations (Fig. 4F). Out of the latter, a concentration-response curve was calculated that revealed an IC_{50} value of 2.23 (1.51–3.30) μM risperidone (Fig. 4G). The Hill coefficient for the respective curve fit was 0.87.

3.4 Risperidone appears more potent in preventing histamine-induced mitochondrial Ca^{2+} signals than cytosolic Ca^{2+} elevations

By comparing the concentration-response curves (Fig. 5A) and IC_{50} values (cytosol: 35.75 μM; mitochondrial matrix: 2.23 μM) for risperidone on cytosolic and mitochondrial Ca^{2+} elevations in response to 100 μM histamine, risperidone appears to be more powerful to prevent mitochondrial Ca^{2+} signals than that in the cytosol. This discrepancy is particularly surprising as mitochondrial Ca^{2+} elevation is fueled by Ca^{2+} increases in the cytosol

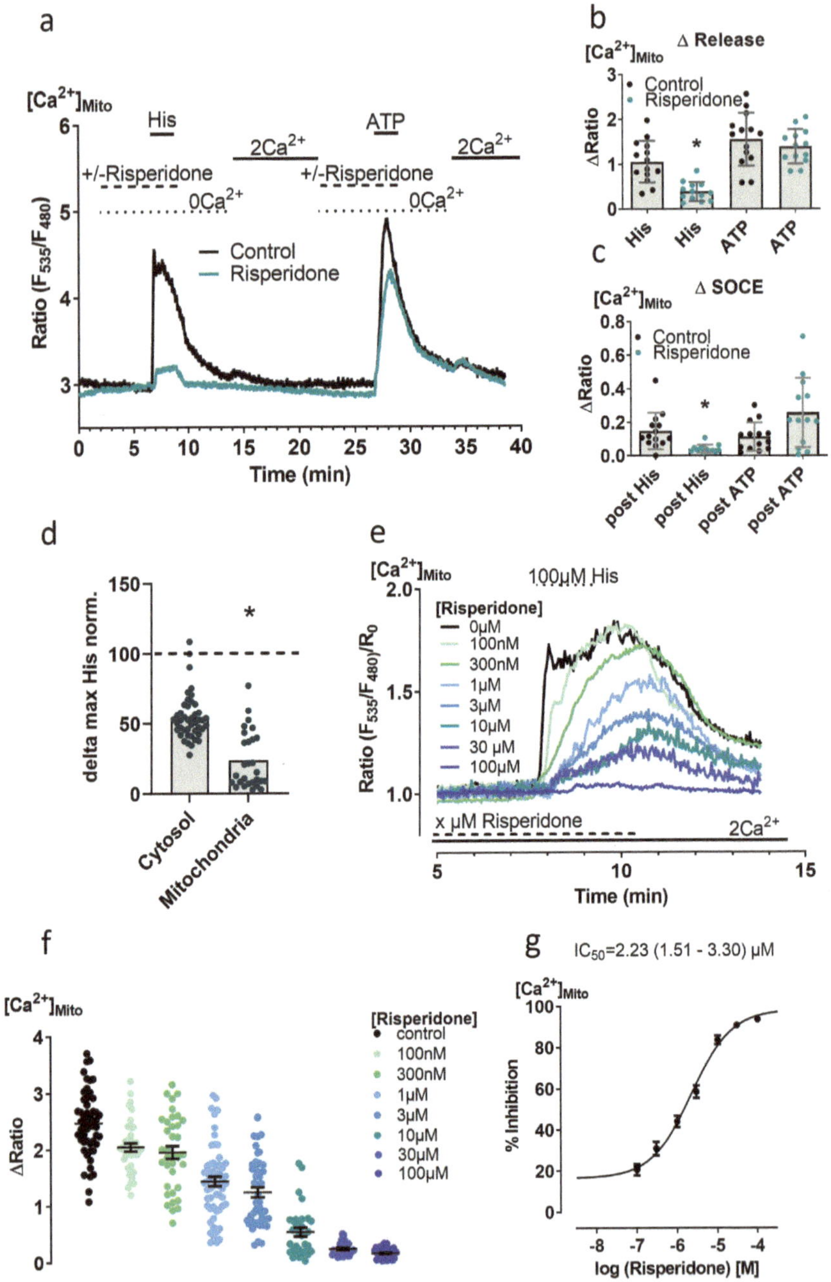

Fig. 4 Effect of risperidone on mitochondrial Ca^{2+} signals in single endothelial cells stimulated with either histamine or ATP. (A) Comparison of the effect of risperidone on mitochondrial Ca^{2+} signals in single endothelial cells serially stimulated with histamine (100 μM) and ATP (100 μM) in the presence of 10 μM risperidone. As indicated cells were stimulated with the respective agonist in the absence of extracellular Ca^{2+} followed by the addition of extracellular Ca^{2+} after agonist washout. Cells were

due to intracellular Ca^{2+} release and SOCE. Notably, as risperidone showed no effect on ATP-induced mitochondrial Ca^{2+} uptake (Fig. 4A–C), a direct effect of risperidone on mitochondrial Ca^{2+} uptake machinery can be excluded. To verify whether these differences build on the different Ca^{2+} affinities of fura-2 (K_D ~200 nM) (Groden, Guan, & Stokes, 1991; Grynkiewicz, Poenie, & Tsien, 1985) and 4mtD3cpv (K_D ~600 nM) (Palmer et al., 2006; Palmer & Tsien, 2006), we expressed the cytosolic D3cpv and compared the inhibitory potency of 10 μM risperidone on a Ca^{2+} signal evoked by 100 μM histamine. Despite the identical K_D values of the D3cpv sensor in the cytosol and the mitochondrial matrix, risperidone was more powerful to prevent histamine-triggered mitochondrial Ca^{2+} signals than that in the cytosol, while no difference in the reduction was found in the cytosol when using fura-2 or D3cpv (Fig. 5B). These data point to a non-linear correlation between cytosolic Ca^{2+} elevations and mitochondrial Ca^{2+} uptake.

3.5 Using various concentrations of risperidone, a linear cytosolic Ca^{2+} elevation upon histamine was achieved

To study the non-linear correlation of cytosolic vs mitochondrial Ca^{2+} levels in more detail, we aimed to correlate defined and robust kinetics of cytosolic Ca^{2+} elevations with increases in the mitochondrial matrix. We assume, that by using an IP_3-generating agonist, the on-kinetics of the cytosolic Ca^{2+} elevation is the main determinant of the transfer of IP_3-released ER Ca^{2+} toward the mitochondrial matrix. If using different histamine concentrations, the on-kinetics in the cytosol generally increased

transiently transfected with a genetically encoded Ca^{2+} sensor 4mtD3cpv. (B and C) Respective statistics, (Control: 9/14; Risperidone: 9/13). (D) Comparison of the normalized Ca^{2+} signals in response to 1 μM histamine in the cytosol (fura-2) and mitochondria (4mtD3cpv) (Cytosol: 6/38; Mitochondria: 9/28). Signals were normalized in regard of the maximal achievable Ca^{2+} signal in response to 100 μM histamine. (E) Representative tracings of mitochondrial Ca^{2+} signals in response of 100 μM histamine in the absence and presence of risperidone (100 nM–100 μM) in single endothelial cells that were transiently transfected with the mitochondrial-targeted Ca^{2+} sensor 4mtD3cpv. (F) Summary of the effect of various risperidone concentrations on histamine-induced mitochondrial Ca^{2+} signaling in 4mtD3cpv expressing single endothelial cells analyzed as the area under the curve (AUC) (concentration in μM risperidone-wells/cells: 0–11/53; 0.1–9/40; 0.3–6/36; 1–9/58; 3–10/47; 10–9/32; 30–9/30; 100–6/34). (G) Concentration-response curve of risperidone on mitochondrial Ca^{2+} elevations (shown in panel F) in single endothelial cells transiently expressing 4mtD3cpv.

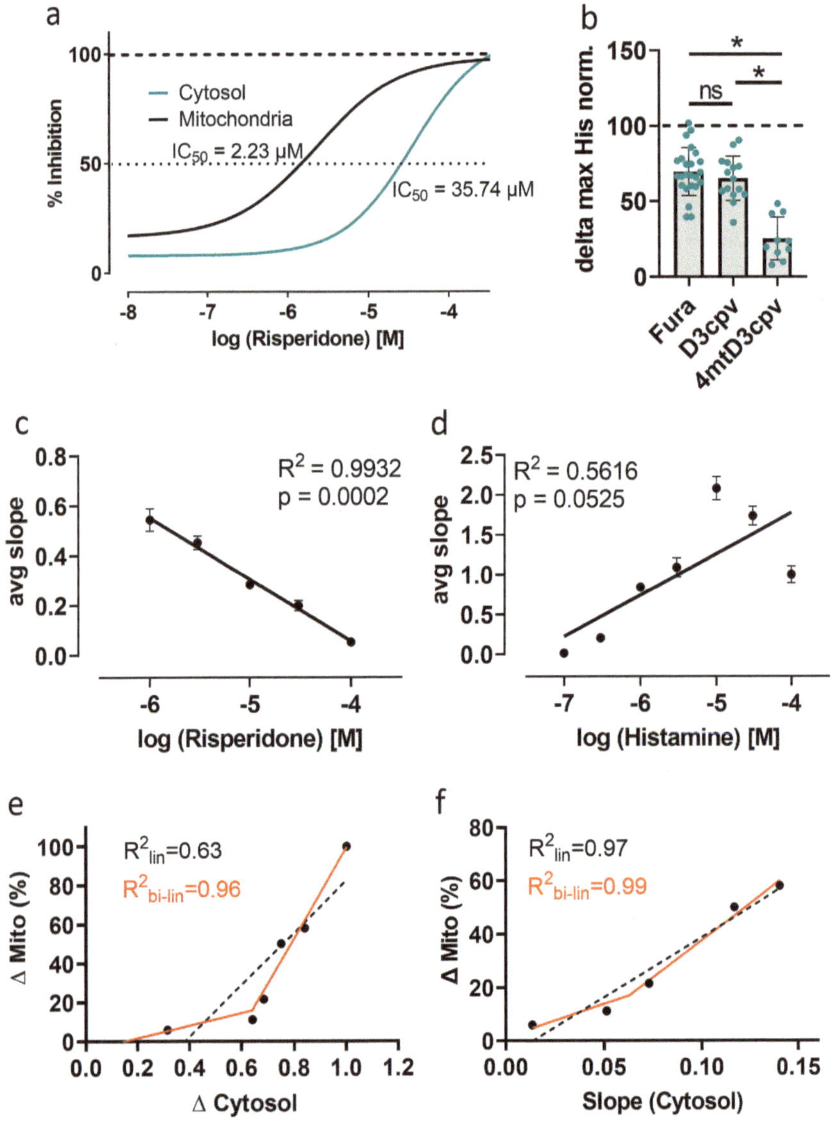

Fig. 5 The correlation between cytosolic and mitochondrial Ca^{2+} elevations in response to an IP_3-triggered ER Ca^{2+} release is non-linear and biphasic. (A) Fitted concentration response curves of the inhibitory effect of risperidone on cytosolic (see Fig. 1C) and mitochondrial Ca^{2+} (3g) elevations in response to ER Ca^{2+} release triggered by 100 μM histamine. (B) Comparison of the effect of 10 μM risperidone on histamine-(100 μM) induced Ca^{2+} signaling in the cytosol and mitochondrial matrix, measured by fura-2, D3cpv, and 4mtD3cpv, respectively (n = cells/wells; Fura-2: 9/23; D3cpv: 9/15; 4mtD3cpv: 9/10). (C) Correlation between the average slope of initial Ca^{2+} elevations in the cytosol achieved by various histamine concentrations (n = wells/cells; 0.1 μM: 5/17; 0.3 μM: 7/44; 1 μM: 9/38; 3 μM: 8/39; 10 μM: 8/55; 100 μM: 6/26).

with increased agonist concentration but did not show a linear concentration dependency in the on-kinetics (slope) (Fig. 5C). Obviously, such an approach would not be suitable to study the relationship between cytosolic Ca^{2+} elevation and mitochondrial Ca^{2+} uptake. However, by using one given histamine concentration (i.e., 100 nM) in the presence of various risperidone concentrations (Fig. 5E), one achieves a linear decrease in the on-kinetics of cytosolic Ca^{2+} elevations with increasing risperidone concentrations (Fig. 5D). Accordingly, this "combined approach" of using one given histamine concentration (i.e., 100 μM) with various concentrations of the histamine receptor antagonist risperidone, appears suitable for studying the correlation between cytosolic Ca^{2+} elevation and mitochondrial Ca^{2+} uptake.

3.6 The correlation between cytosolic and mitochondrial Ca^{2+} elevations is biphasic but linear after a certain threshold

Using the "combined approach" we were now able to correlate the Ca^{2+} elevations in the mitochondrial matrix (ΔMito) with the respective cytosolic Ca^{2+} elevations (ΔCytosol) (Fig. 5E), and their on-kinetics (Slope Cytosol) (Fig. 5F). Both analyses revealed a biphasic correlation with a non-responding phase at lower cytosolic Ca^{2+} elevations and a linear correlating responding phase in which cytosolic Ca^{2+} elevations correspond well with the respective matrix Ca^{2+} increases.

4. Discussion

Mitochondria are actively participating in cellular Ca^{2+} homeostasis. Recently, many phenomena of the mitochondrial Ca^{2+} homeostasis could

Parameters and equations for the fit for linear correlation are given in the graph. (D) Correlation between the average slope of initial Ca^{2+} elevations upon 100 μM histamine in the presence of various risperidone concentrations. Data are extracted from that presented in Fig. 1 and equation/parameters for the fit for linear correlation are given in the graph. (E and F) Correlation of the initial mitochondrial Ca^{2+} elevations normalized to the maximal increase achieved with the respective initial cytosolic Ca^{2+} elevations (E) and initial slope to maximum of cytosolic Ca^{2+} elevations (F) to 100 μM histamine in the presence of 1–100 μM risperidone. Data were extracted from the experiments shown in Fig. 3E and F. Non-linear biphasic fit (red) and linear fits for the two phases (black) are shown.

be molecularly attributed to newly discovered proteins, like those of the MCUC (Feno et al., 2020; Mammucari et al., 2017; Marchi & Pinton, 2014) or NCLX as the main route of mitochondrial Ca^{2+} extrusion (Marchi et al., 2014; Palty et al., 2010). Nevertheless, despite all this progress in our understanding of how mitochondrial Ca^{2+} homeostasis is organized, there are plenty of open issues that await an answer.

One aspect that is still unresolved, is how Ca^{2+} that enters the cell via plasma membrane ion channels, like the SOCE pathway, is sequestered by the mitochondria. Tulio Pozzan and his coworker firstly described that intracellular Ca^{2+} release from the ER generates Ca^{2+} hot spots of higher than 10 μM Ca^{2+} (Giacomello et al., 2010), thus matching the Ca^{2+} sensitivity of MICU1 (Waldeck-Weiermair et al., 2015), the gatekeeper of the MCUC (Ahuja & Muallem, 2014; Foskett & Madesh, 2014; Kamer & Mootha, 2014; Plovanich et al., 2013). However, surface Ca^{2+} elevations upon SOCE activation are too small (Giacomello et al., 2010) to activate MICU1. We have reported that Letm1, a protein that was suggested to serve as mitochondrial Ca^{2+}/H^{+} antiporter (Jiang, Zhao, & Clapham, 2009), is exclusively involved in mitochondrial Ca^{2+} uptake of Ca^{2+} entering the cell via the SOCE but not that released from the ER (Doonan et al., 2014; Waldeck-Weiermair et al., 2011, 2013). Thus, alternative routes for Ca^{2+} to enter the mitochondria pending the source, concentration, and/or kinetics by which Ca^{2+} diffuses to the mitochondrial surface may exist. Alternatively, posttranslational modification of the proteins that belong to the MCUC might allow Ca^{2+} influx into the mitochondria, even below the Ca^{2+} threshold of MICU1. Excitingly, the group of Muniswamy Madesh has described that oxidation of MCU yields activation of the MCU and its translocation into the inner boundary membrane (Dong et al., 2017; Nemani, Shanmughapriya, & Madesh, 2018), a process that appears important for MCUC reassembly and activity in the inner boundary membrane (Gottschalk et al., 2019; Gottschalk, Madreiter-Sokolowski, & Graier, 2021). The activity of MCUC can also be attenuated by arginine methylation of MICU1 by the protein arginine methyltransferase 1 that yields decreased Ca^{2+} binding affinity and, thus, reduced susceptibility for Ca^{2+}-mediated activation upon intracellular Ca^{2+} release (Madreiter-Sokolowski et al., 2016). Notably, UCP2, which was found to be fundamental for mitochondrial Ca^{2+} uptake under certain conditions (Trenker, Malli, Fertschai, Levak-Frank, & Graier, 2007), normalizes the

Ca^{2+} sensitivity of methylated MICU1 and, consequently, re-establishes normal mitochondrial Ca^{2+} uptake in cells with high PRMT1 activity (Madreiter-Sokolowski et al., 2016). Moreover, the activity of the MCUC also critically depends on the stoichiometry of the MICU1 isomers (Patron et al., 2018; Payne, Hoff, Roskowski, & Foskett, 2017; Plovanich et al., 2013; Wu et al., 2019, 2020). All these studies clearly indicate that despite the enormous progress during recent years, the phenomenon of mitochondrial Ca^{2+} uptake is not fully understood today.

The very simple approach of performing a concentration-response curve of a receptor antagonist risperidone on histamine-evoked cytosolic and mitochondrial Ca^{2+} signaling revealed an approximately 15 times higher obvious sensitivity of mitochondrial Ca^{2+} uptake in response to histamine to risperidone than the respective cytosolic Ca^{2+} signal. These findings are surprising considering that the mitochondrial Ca^{2+} uptake is fueled by cytosolic Ca^{2+} elevations, thus, one would expect similar IC_{50} values for risperidone for the Ca^{2+} signal in both compartments. Hence, our data that risperidone did not affect ATP-induced cytosolic or mitochondrial Ca^{2+} signaling, excludes a further effect of risperidone on, e.g., the mitochondrial Ca^{2+} uptake machinery besides H_1 receptor inhibition. Moreover, by using the identical genetically encoded biosensor, D3cpv without the mitochondrial targeting sequence for monitoring cytosolic Ca^{2+}, we found very similar sensitivities in the cytosolic Ca^{2+} signal to risperidone to that of fura-2. These data let us rule out that the different K_Ds of fura-2 and 4mtD3cpv for Ca^{2+} are causing the great difference in the sensitivity to risperidone between the cytosolic and mitochondrial Ca^{2+} signals.

Our data discussed above point to a non-linear correlation between cytosolic and mitochondrial Ca^{2+} elevations. Therefore, we searched for a simple approach that would allow us to control the cytosolic Ca^{2+} concentration without disturbing the inter-organellar network as it would happen using ionomycin or digitonin for clamping cytosolic Ca^{2+} levels. Our data presented in this work demonstrated a selective, and strictly concentration-dependent attenuation of histamine-induced cytosolic Ca^{2+} elevations by ER Ca^{2+} release and SOCE in single endothelial cells by risperidone, most likely by its inhibition of the H_1 receptor (Leysen et al., 1988, 1992) found in endothelial cells (Spitaler et al., 2002; Voorde & Leusen, 1983). Because we combined one given concentration of an agonist with various concentrations of its receptor antagonist to achieve well-defined cytosolic Ca^{2+} signals, we named this approach for titrating cytosolic Ca^{2+} elevations "combined approach." Because the molecular pathway(s) involved in the mitochondrial uptake of Ca^{2+} that enters the cells via the SOCE is still not entirely resolved, we concentrated on the initial phase of the histamine-induced Ca^{2+} signaling, which is due to ER Ca^{2+} release that is known to be transferred into the mitochondria via the MCUC. Using the "combined approach" we could achieve a linear change in the on-kinetics of the cytosolic Ca^{2+} elevation (slope). In contrast, using various concentrations of the agonist alone showed a multiphasic, non-linear correlation in the on-kinetics of the cytosolic Ca^{2+} elevation, and thus was not suitable to study the correlation of initial cytosolic Ca^{2+} elevation with mitochondrial Ca^{2+} uptake.

Applying our "combined approach," the correlation between cytosolic Ca^{2+} elevations and mitochondrial Ca^{2+} signals showed a biphasic correlation. Our data indicate that at lower cytosolic Ca^{2+} elevations, the cytosolic Ca^{2+} appears not to be transferred into the mitochondria. However, at a certain cytosolic Ca^{2+} level, a linear correlation between cytosolic and mitochondrial Ca^{2+} signaling was observed. Notably, the correlation between the on-kinetics of cytosolic Ca^{2+} elevations that may reflect the kinetics of the Ca^{2+} floating toward the mitochondrial surface did not provide such a clear biphasic correlation. These data may indicate that for the correlation between cytosolic Ca^{2+} elevations and mitochondrial Ca^{2+} uptake the ion concentration, but not the kinetics of Ca^{2+} at the mitochondrial surface, is the main determinant. These findings support previous work indicating a threshold of approximately 5–6 μM Ca^{2+} to overcome the inhibitory function of MICU1 by triggering its rearrangement (Waldeck-Weiermair et al., 2015; Wang et al., 2014). Hence, this threshold may explain the necessity of alternative regulation or pathways for mitochondrial sequestration of Ca^{2+} that enters the cell via SOCE.

In the present work, we have established a simple experimental approach to study the correlation between cytosolic and mitochondrial Ca^{2+} signals in response to IP_3-triggered ER Ca^{2+} release. Our data support the concept of a threshold of cytosolic Ca^{2+} (at the mitochondrial surface) (Fig. 6). Below this threshold level, no mitochondrial Ca^{2+} uptake occurs while above it, a linear correlation between cytosolic and mitochondrial Ca^{2+} elevations exists. These data strongly support the exquisitely specific interaction of the ER and mitochondria in endothelial Ca^{2+} regulation.

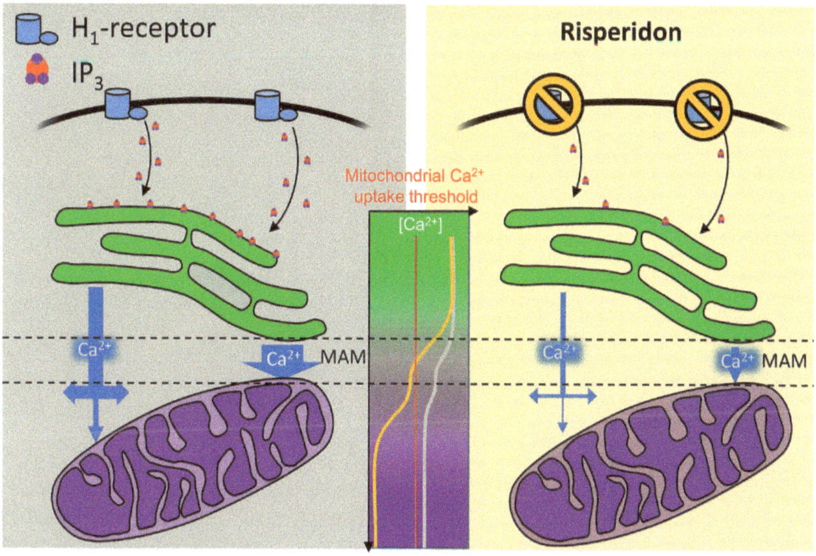

Fig. 6 Schematic illustrating the potentiated effect of a reduction of ER Ca^{2+} release by risperidone on mitochondrial Ca^{2+} uptake. ER Ca^{2+} release leads to high cytosolic Ca^{2+} concentration at mitochondrial-associated ER membrane (MAM) structures. These Ca^{2+} hotspots surpass a Ca^{2+} concentration threshold leading to mitochondrial Ca^{2+} uptake. In the case of reduced ER Ca^{2+} release due to the partial inhibition of histamine receptors by risperidone, MAM Ca^{2+} concentrations do not exceed the mitochondrial Ca^{2+} uptake threshold. Consequently, the obvious inhibitory effect of risperidone is more pronounced on the mitochondrial Ca^{2+} uptake than that on the cytosolic Ca^{2+} signaling despite the compound does not directly affect mitochondrial Ca^{2+} uptake machinery.

Conflict of interest

The authors declare that they have no conflict of interest.

Acknowledgments

The authors wish to thank Anna Schreilechner, BSc, for their excellent technical assistance and Prof. Dr. C. J. S. Edgell (University of North Carolina, Chapel Hill, NC, USA) for providing us the EA.hy926 cells. This work was supported by the Austrian Science Fund (FWF) (DK-MCD W1226 to W.F.G), the MEFO Graz (to W.F.G.), and Nikon Austria (to W.F.G.). O.A.B. are doctoral fellows within the doctoral program Metabolic and Cardiovascular Disease (MCD) (FWF, DKplus W 1226-B18) at the Medical University of Graz. The SIM equipment is part of the Nikon-Center of Excellence, Graz, that is supported by the Austrian infrastructure program 2013/2014, Nikon Austria Inc. and BioTechMed.

Author contributions

O.A.B. performed the experimental work. O.A.B. and B.G. performed data analyses. W.F.G. and R.M. supervised the research and project planning, performed data interpretation. W.F.G. and O.A.B. prepared the manuscript. All authors discussed the results and implications and commented on the manuscript.

References

Ahuja, M., & Muallem, S. (2014). The gatekeepers of mitochondrial calcium influx: MICU1 and MICU2. *EMBO Reports*, *15*, 205–206. https://doi.org/10.1002/embr.201438446.

Baughman, J. M., Perocchi, F., Girgis, H. S., Plovanich, M., Belcher-Timme, C. A., Sancak, Y., et al. (2011). Integrative genomics identifies MCU as an essential component of the mitochondrial calcium uniporter. *Nature*, *476*, 341. https://doi.org/10.1038/nature10234.

Csordás, G., Golenár, T., Seifert, E. L., Kamer, K. J., Sancak, Y., Perocchi, F., et al. (2013). MICU1 controls both the threshold and cooperative activation of the mitochondrial Ca^{2+} uniporter. *Cell Metabolism*, *17*, 976–987. https://doi.org/10.1016/j.cmet.2013.04.020.

Dong, Z., Shanmughapriya, S., Tomar, D., Siddiqui, N., Lynch, S., Nemani, N., et al. (2017). Mitochondrial Ca^{2+} uniporter is a mitochondrial luminal redox sensor that augments MCU channel activity. *Molecular Cell*, *65*, 1014–1028.e7. https://doi.org/10.1016/j.molcel.2017.01.032.

Doonan, P. J., Chandramoorthy, H. C., Hoffman, N. E., Zhang, X., Cárdenas, C., Shanmughapriya, S., et al. (2014). LETM1-dependent mitochondrial Ca^{2+} flux modulates cellular bioenergetics and proliferation. *The FASEB Journal*, *28*, 4936–4949. https://doi.org/10.1096/fj.14-256453.

Edgell, C. J., McDonald, C. C., & Graham, J. B. (1983). Permanent cell line expressing human factor VIII-related antigen established by hybridization. *Proceedings of the National Academy of Sciences of the United States of America*, *80*, 3734–3737. https://doi.org/10.1073/pnas.80.12.3734.

Feno, S., Rizzuto, R., Raffaello, A., & Reane, D. V. (2020). The molecular complexity of the mitochondrial calcium uniporter. *Cell Calcium*, *93*, 102322. https://doi.org/10.1016/j.ceca.2020.102322.

Foskett, J. K., & Madesh, M. (2014). Regulation of the mitochondrial Ca^{2+} uniporter by MICU1 and MICU2. *Biochemical and Biophysical Research Communications*, *449*, 377–383. https://doi.org/10.1016/j.bbrc.2014.04.146.

Giacomello, M., Drago, I., Bortolozzi, M., Scorzeto, M., Gianelle, A., Pizzo, P., et al. (2010). Ca^{2+} hot spots on the mitochondrial surface are generated by Ca^{2+} mobilization from stores, but not by activation of store-operated Ca^{2+} channels. *Molecular Cell*, *38*, 280–290. https://doi.org/10.1016/j.molcel.2010.04.003.

Gottschalk, B., Klec, C., Leitinger, G., Bernhart, E., Rost, R., Bischof, H., et al. (2019). MICU1 controls cristae junction and spatially anchors mitochondrial Ca^{2+} uniporter complex. *Nature Communications*, *10*, 3732. https://doi.org/10.1038/s41467-019-11692-x.

Gottschalk, B., Madreiter-Sokolowski, C. T., & Graier, W. F. (2021). Cristae junction as a fundamental switchboard for mitochondrial ion signaling and bioenergetics. *Cell Calcium*, *101*, 102517. https://doi.org/10.1016/j.ceca.2021.102517.

Graier, W. F., Kurt, S., & Kukovetz, W. R. (1990). Effect of sodium fluoride on cytosolic free Ca^{2+}-concentrations and cGMP-levels in endothelial cells. *Cellular Signalling*, *2*, 369–375. https://doi.org/10.1016/0898-6568(90)90067-k.

Graier, W. F., Simecek, S., & Sturek, M. (1995). Cytochrome P450 mono-oxygenase-regulated signalling of Ca^{2+} entry in human and bovine endothelial cells. *The Journal of Physiology, 482*, 259–274. https://doi.org/10.1113/jphysiol.1995.sp020515.

Groden, D. L., Guan, Z., & Stokes, B. T. (1991). Determination of Fura-2 dissociation constants following adjustment of the apparent Ca-EGTA association constant for temperature and ionic strength. *Cell Calcium, 12*, 279–287. https://doi.org/10.1016/0143-4160(91)90002-v.

Grynkiewicz, G., Poenie, M., & Tsien, R. Y. (1985). A new generation of Ca^{2+} indicators with greatly improved fluorescence properties. *The Journal of Biological Chemistry, 260*, 3440–3450. https://doi.org/10.1016/s0021-9258(19)83641-4.

Jiang, D., Zhao, L., & Clapham, D. E. (2009). Genome-wide RNAi screen identifies Letm1 as a mitochondrial Ca^{2+}/H^{+} antiporter. *Science, 326*, 144–147. https://doi.org/10.1126/science.1175145.

Kamer, K. J., & Mootha, V. K. (2014). MICU1 and MICU2 play nonredundant roles in the regulation of the mitochondrial calcium uniporter. *EMBO Reports, 15*, 299–307. https://doi.org/10.1002/embr.201337946.

Koshenov, Z., Oflaz, F. E., Hirtl, M., Gottschalk, B., Rost, R., Malli, R., et al. (2022). Citrin mediated metabolic rewiring in response to altered basal subcellular Ca^{2+} homeostasis. *Communications Biology, 5*, 76. https://doi.org/10.1038/s42003-022-03019-2.

Leysen, J. E., Gommeren, W., Eens, A., de Chaffoy de Courcelles, D., Stoof, J. C., & Janssen, P. A. (1988). Biochemical profile of risperidone, a new antipsychotic. *The Journal of Pharmacology and Experimental Therapeutics, 247*, 661–670.

Leysen, J. E., Janssen, P. M., Gommeren, W., Wynants, J., Pauwels, P. J., & Janssen, P. A. (1992). In vitro and in vivo receptor binding and effects on monoamine turnover in rat brain regions of the novel antipsychotics risperidone and ocaperidone. *Molecular Pharmacology, 41*, 494–508.

Madreiter-Sokolowski, C. T., Klec, C., Parichatikanond, W., Stryeck, S., Gottschalk, B., Pulido, S., et al. (2016). PRMT1-mediated methylation of MICU1 determines the UCP2/3 dependency of mitochondrial Ca^{2+} uptake in immortalized cells. *Nature Communications*, 7. https://doi.org/10.1038/ncomms12897. ncomms12897.

Malli, R., Frieden, M., Hunkova, M., Trenker, M., & Graier, W. F. (2007). Ca^{2+} refilling of the endoplasmic reticulum is largely preserved albeit reduced Ca^{2+} entry in endothelial cells. *Cell Calcium, 41*, 63–76. https://doi.org/10.1016/j.ceca.2006.05.001.

Mallilankaraman, K., Cárdenas, C., Doonan, P. J., Chandramoorthy, H. C., Irrinki, K. M., Golenár, T., et al. (2012). MCUR1 is an essential component of mitochondrial Ca^{2+} uptake that regulates cellular metabolism. *Nature Cell Biology, 14*, 1336. https://doi.org/10.1038/ncb2622.

Mammucari, C., Gherardi, G., & Rizzuto, R. (2017). Structure, activity regulation, and role of the mitochondrial calcium uniporter in health and disease. *Frontiers in Oncology*, 7, 139. https://doi.org/10.3389/fonc.2017.00139.

Marchi, S., & Pinton, P. (2014). The mitochondrial calcium uniporter complex: Molecular components, structure and physiopathological implications. *The Journal of Physiology, 592*, 829–839. https://doi.org/10.1113/jphysiol.2013.268235.

Marchi, U. D., Santo-Domingo, J., Castelbou, C., Sekler, I., Wiederkehr, A., & Demaurex, N. (2014). NCLX protein, but not LETM1, mediates mitochondrial Ca^{2+} extrusion, thereby limiting Ca^{2+}-induced NAD(P)H production and modulating matrix redox state. *The Journal of Biological Chemistry, 289*, 20377–20385. https://doi.org/10.1074/jbc.m113.540898.

Modesti, L., Danese, A., Vitto, V. A. M., Ramaccini, D., Aguiari, G., Gafà, R., et al. (2021). Mitochondrial Ca^{2+} signaling in health, disease and therapy. *Cells, 10*, 1317. https://doi.org/10.3390/cells10061317.

Nemani, N., Shanmughapriya, S., & Madesh, M. (2018). Molecular regulation of MCU: Implications in physiology and disease. *Cell Calcium, 74*, 86–93. https://doi.org/10.1016/j.ceca.2018.06.006.

Oflaz, F. E., Koshenov, Z., Hirtl, M., Rost, R., Bachkoenig, O. A., Gottschalk, B., et al. (2021). Near-UV light induced ROS production initiates spatial Ca^{2+} spiking to fire NFATc3 translocation. *International Journal of Molecular Sciences, 22*, 8189. https://doi.org/10.3390/ijms22158189.

Palmer, A. E., Giacomello, M., Kortemme, T., Hires, S. A., Lev-Ram, V., Baker, D., et al. (2006). Ca^{2+} indicators based on computationally redesigned calmodulin-peptide pairs. *Chemistry & Biology, 13*, 521–530. https://doi.org/10.1016/j.chembiol.2006.03.007.

Palmer, A. E., & Tsien, R. Y. (2006). Measuring calcium signaling using genetically targetable fluorescent indicators. *Nature Protocols, 1*, 1057. https://doi.org/10.1038/nprot.2006.172.

Palty, R., Silverman, W. F., Hershfinkel, M., Caporale, T., Sensi, S. L., Parnis, J., et al. (2010). NCLX is an essential component of mitochondrial Na^{+}/Ca^{2+} exchange. *Proceedings of the National Academy of Sciences of the United States of America, 107*, 436–441. https://doi.org/10.1073/pnas.0908099107.

Patron, M., Granatiero, V., Espino, J., Rizzuto, R., & Stefani, D. D. (2018). MICU3 is a tissue-specific enhancer of mitochondrial calcium uptake. *Cell Death and Differentiation, 26*, 179–195. https://doi.org/10.1038/s41418-018-0113-8.

Payne, R., Hoff, H., Roskowski, A., & Foskett, J. K. (2017). MICU2 restricts spatial crosstalk between InsP3R and MCU channels by regulating threshold and gain of MICU1-mediated inhibition and activation of MCU. *Cell Reports, 21*, 3141–3154. https://doi.org/10.1016/j.celrep.2017.11.064.

Perocchi, F., Gohil, V. M., Girgis, H. S., Bao, X. R., McCombs, J. E., Palmer, A. E., et al. (2010). MICU1 encodes a mitochondrial EF hand protein required for Ca^{2+} uptake. *Nature, 467*, 291. https://doi.org/10.1038/nature09358.

Plovanich, M., Bogorad, R. L., Sancak, Y., Kamer, K. J., Strittmatter, L., Li, A. A., et al. (2013). MICU2, a paralog of MICU1, resides within the mitochondrial uniporter complex to regulate calcium handling. *PLoS One, 8*, e55785. https://doi.org/10.1371/journal.pone.0055785.

Pozzan, T., & Rizzuto, R. (2000). High tide of calcium in mitochondria. *Nature Cell Biology, 2*. https://doi.org/10.1038/35000095. ncb0200_E25.

Puente, B. N., Sun, J., Parks, R. J., Fergusson, M. M., Liu, C., Springer, D., et al. (2020). MICU3 plays an important role in cardiovascular function. *Circulation Research, 127*, 1571–1573. https://doi.org/10.1161/circresaha.120.317177.

Rizzuto, R., Simpson, A. W. M., Brini, M., & Pozzan, T. (1992). Rapid changes of mitochondrial Ca^{2+} revealed by specifically targeted recombinant aequorin. *Nature, 358*, 358325a0. https://doi.org/10.1038/358325a0.

Sancak, Y., Markhard, A. L., Kitami, T., Kovács-Bogdán, E., Kamer, K. J., Udeshi, N. D., et al. (2013). EMRE is an essential component of the mitochondrial calcium uniporter complex. *Science, 342*, 1379–1382. https://doi.org/10.1126/science.1242993.

Spitaler, M. M., Hammer, A., Malli, R., & Graier, W. F. (2002). Functional analysis of histamine receptor subtypes involved in endothelium-mediated relaxation of the human uterine artery. *Clinical and Experimental Pharmacology & Physiology, 29*, 711–716. https://doi.org/10.1046/j.1440-1681.2002.03704.x.

Stefani, D. D., Raffaello, A., Teardo, E., Szabò, I., & Rizzuto, R. (2011). A forty-kilodalton protein of the inner membrane is the mitochondrial calcium uniporter. *Nature, 476*, 336. https://doi.org/10.1038/nature10230.

Trenker, M., Malli, R., Fertschai, I., Levak-Frank, S., & Graier, W. F. (2007). Uncoupling proteins 2 and 3 are fundamental for mitochondrial Ca^{2+} uniport. *Nature Cell Biology, 9*, 445–452. https://doi.org/10.1038/ncb1556.

Voorde, J. V. D., & Leusen, I. (1983). Role of the endothelium in the vasodilator response of rat thoracic aorta to histamine. *European Journal of Pharmacology*, *87*, 113–120. https://doi.org/10.1016/0014-2999(83)90056-0.

Waldeck-Weiermair, M., Alam, M. R., Khan, M. J., Deak, A. T., Vishnu, N., Karsten, F., et al. (2012). Spatiotemporal correlations between cytosolic and mitochondrial Ca^{2+} signals using a novel red-shifted mitochondrial targeted cameleon. *PLoS One*, *7*, e45917. https://doi.org/10.1371/journal.pone.0045917.

Waldeck-Weiermair, M., Deak, A. T., Groschner, L. N., Alam, M. R., Jean-Quartier, C., Malli, R., et al. (2013). Molecularly distinct routes of mitochondrial Ca^{2+} uptake are activated depending on the activity of the sarco/endoplasmic reticulum Ca^{2+} ATPase (SERCA). *The Journal of Biological Chemistry*, *288*, 15367–15379. https://doi.org/10.1074/jbc.m113.462259.

Waldeck-Weiermair, M., Duan, X., Naghdi, S., Khan, M. J., Trenker, M., Malli, R., et al. (2010). Uncoupling protein 3 adjusts mitochondrial Ca^{2+} uptake to high and low Ca^{2+} signals. *Cell Calcium*, *48*, 288–301. https://doi.org/10.1016/j.ceca.2010.10.004.

Waldeck-Weiermair, M., Gottschalk, B., Madreiter-Sokolowski, C. T., Ramadani-Muja, J., Ziomek, G., Klec, C., et al. (2019). Development and application of sub-mitochondrial targeted Ca^{2+} biosensors. *Frontiers in Cellular Neuroscience*, *13*, 449. https://doi.org/10.3389/fncel.2019.00449.

Waldeck-Weiermair, M., Jean-Quartier, C., Rost, R., Khan, M. J., Vishnu, N., Bondarenko, A. I., et al. (2011). Leucine zipper EF hand-containing transmembrane protein 1 (Letm1) and uncoupling proteins 2 and 3 (UCP2/3) contribute to two distinct mitochondrial Ca^{2+} uptake pathways. *The Journal of Biological Chemistry*, *286*, 28444–28455. https://doi.org/10.1074/jbc.m111.244517.

Waldeck-Weiermair, M., Malli, R., Naghdi, S., Trenker, M., Kahn, M. J., & Graier, W. F. (2010). The contribution of UCP2 and UCP3 to mitochondrial Ca^{2+} uptake is differentially determined by the source of supplied Ca^{2+}. *Cell Calcium*, *47*, 433–440. https://doi.org/10.1016/j.ceca.2010.03.004.

Waldeck-Weiermair, M., Malli, R., Parichatikanond, W., Gottschalk, B., Madreiter-Sokolowski, C. T., Klec, C., et al. (2015). Rearrangement of MICU1 multimers for activation of MCU is solely controlled by cytosolic Ca^{2+}. *Scientific Reports*, *5*, 15602. https://doi.org/10.1038/srep15602.

Wang, L., Yang, X., Li, S., Wang, Z., Liu, Y., Feng, J., et al. (2014). Structural and mechanistic insights into MICU1 regulation of mitochondrial calcium uptake. *The EMBO Journal*, *33*, 594–604. https://doi.org/10.1002/embj.201386523.

Wu, W., Shen, Q., Lei, Z., Qiu, Z., Li, D., Pei, H., et al. (2019). The crystal structure of MICU2 provides insight into Ca^{2+} binding and MICU1-MICU2 heterodimer formation. *EMBO Reports*, *20*, e47488. https://doi.org/10.15252/embr.201847488.

Wu, W., Shen, Q., Zhang, R., Qiu, Z., Wang, Y., Zheng, J., et al. (2020). The structure of the MICU1-MICU2 complex unveils the regulation of the mitochondrial calcium uniporter. *The EMBO Journal*, *39*, e104285. https://doi.org/10.15252/embj.2019104285.

CHAPTER THREE

Regulation of exosome release by lysosomal acid ceramidase in coronary arterial endothelial cells: Role of TRPML1 channel

Guangbi Li[a], Dandan Huang[a], Pengyang Li[b], Xinxu Yuan[a], Viktor Yarotskyy[a], and Pin-Lan Li[a,*]

[a]Department of Pharmacology and Toxicology, School of Medicine, Virginia Commonwealth University, Richmond, VA, United States
[b]Division of Cardiology, School of Medicine, Virginia Commonwealth University, Richmond, VA, United States
*Corresponding author: e-mail address: pin-lan.li@vcuhealth.org

Contents

Current Topics in Membranes, Volume 90
ISSN 1063-5823
https://doi.org/10.1016/bs.ctm.2022.09.002

Abstract

Lysosomal acid ceramidase (AC) has been reported to determine multivesicular body (MVB) fate and exosome secretion in different mammalian cells including coronary arterial endothelial cells (CAECs). However, this AC-mediated regulation of exosome release from CAECs and associated underlying mechanism remain poorly understood. In the present study, we hypothesized that AC controls lysosomal Ca^{2+} release through TRPML1 channel to regulate exosome release in murine CAECs. To test this hypothesis, we isolated and cultured CAECs from WT/WT and endothelial cell-specific Asah1 gene (gene encoding AC) knockout mice. Using these CAECs, we first demonstrated a remarkable increase in exosome secretion and significant reduction of lysosome-MVB interaction in CAECs lacking Asah1 gene compared to those cells from WT/WT mice. ML-SA1, a TRPML1 channel agonist, was found to enhance lysosome trafficking and increase lysosome-MVB interaction in WT/WT CAECs, but not in CAECs lacking Asah1 gene. However, sphingosine, an AC-derived sphingolipid, was able to increase lysosome movement and lysosome-MVB interaction in CAECs lacking Asah1 gene, leading to reduced exosome release from these cells. Moreover, Asah1 gene deletion was shown to substantially inhibit lysosomal Ca^{2+} release through suppression of TRPML1 channel activity in CAECs. Sphingosine as an AC product rescued the function of TRPML1 channel in CAECs lacking Asah1 gene. These results suggest that Asah1 gene defect and associated deficiency of AC activity may inhibit TRPML1 channel activity, thereby reducing MVB degradation by lysosome and increasing exosome release from CAECs. This enhanced exosome release from CAECs may contribute to the development of coronary arterial disease under pathological conditions.

1. Introduction

Since the late 1960s, numerous studies have shown that extracellular vesicles (EVs) originate from different subcellular membrane compartments and are released into the interstitial space. After many debates about the characterization and classification of EVs, a general consensus has been recently reached to describe three distinct populations of EVs, including apoptotic bodies, microvesicles, and exosomes, and most cell types in humans can release different vesicles into the interstitial space (Crescitelli et al., 2013; Lazaro-Ibanez et al., 2014). Exosomes, the smallest EV with approximately 40–140 nm in diameter, are formed through the endocytic process and released from intracellular multivesicular bodies (MVBs) through an active process (Rajagopal & Harikumar, 2018). EVs or exosomes have been extensively studied for their biogenesis and related function in cell-to-cell communication and in the pathogenesis of different diseases including cardiovascular diseases (Boulanger, Loyer, Rautou, & Amabile, 2017; Chistiakov, Orekhov, & Bobryshev, 2015; Hessvik & Llorente, 2018;

Perrotta & Aquila, 2016). There are numerous studies about endothelial cell (EC)-derived exosomes as a biomarker for endothelial function or injuries in different cardiovascular or systemic diseases such as diabetes mellitus, systemic lupus erythematosus, and hyperhomocysteinemia. These circulating exosomes may also be an interesting prognostic marker to identify patients at risk of major adverse cardiovascular events (Nozaki et al., 2009; Sinning et al., 2011). Moreover, evidence is accumulating that EVs including exosomes are critical to transport, secret or release various mediators such as ET-1, PGs, microRNAs, bioactive lipids, IL-1β, and even enzymes such as NOX, cyclooxygenases, and caspase-1 from ECs (Davenport et al., 2016; Deng et al., 2018; Feng et al., 2019; Ju et al., 2014; Subra et al., 2010; Yuan, Bhat, Lohner, Zhang, & Li, 2019). However, the regulatory mechanisms of exosome release from ECs under physiological and pathological conditions remain poorly understood.

Based on findings in previous studies, lysosome function may determine the MVB fate through lysosome-dependent degradation of MVBs (Boulanger et al., 2017; Chistiakov et al., 2015; Futter, Pearse, Hewlett, & Hopkins, 1996; Hessvik & Llorente, 2018). It was shown that inhibition of lysosome function with different alkaline agents or lysosomal v-ATPase inhibitor bafilomycin A or chloroquine increased exosome secretion in different cells such as neurons, epithelial cells, and vascular cells (Alvarez-Erviti et al., 2011; Vingtdeux et al., 2007; Wei et al., 2013). Mutations of VPS4 gene required for MVB maturation and fusion with lysosome increased the levels of EV-associated proteins within cells (Hasegawa et al., 2011). In addition, there is evidence that MVBs may be fused with autophagosomes to form amphisomes and subsequently fuse with lysosomes to terminate MVB fate (Baixauli, Lopez-Otin, & Mittelbrunn, 2014; Fader, Sanchez, Furlan, & Colombo, 2008; Murrow & Debnath, 2015). However, most of these previous studies were based on an assumption that lysosomes are passively receiving MVBs to fuse for degradation of these vesicles or their content. In fact, there is considerable evidence that lysosomes have trafficking function within different cells (Bhat, Li, et al., 2020; Bhat, Yuan, Cain, Salloum, & Li, 2020; Bhat, Yuan, Camus, Salloum, & Li, 2020; Yuan et al., 2019). In this regard, our recent studies have demonstrated that acid ceramidase (AC) controls lysosomal Ca^{2+} release through transient receptor potential mucolipin 1 (TRPML1) channel to regulate lysosome trafficking in different types of cells (Bhat, Li, et al., 2020; Li, Huang, et al., 2020; Li et al., 2019). Together, these findings raised the possibility that lysosomes may actively traffic and fuse to MVBs to determine MVB fate and exosome secretion in ECs. Meanwhile, AC activity and TRPML1 channel-mediated Ca^{2+} release may be importantly involved in the lysosomal regulation of exosome release in ECs.

In the present study, we hypothesized that AC controls lysosomal Ca^{2+} release through TRPML1 channel to regulate exosome release in murine coronary arterial endothelial cells (CAECs). To test this hypothesis, we examined whether exosome release from CAECs is enhanced by Asah1 gene knockout. Also, we determined whether Asah1 gene deletion affects lysosome trafficking and lysosome-MVB interaction in CAECs. Moreover, we examined the effects of Asah1 gene deletion and AC-associated sphingolipids on TRPML1 channel activity in these cells. Finally, we tested whether dynein is involved in the lysosomal regulation of exosome release in CAECs. Our results suggest that AC controls lysosome trafficking and lysosome-MVB interaction to determine MVB fate and exosome release in CAECs through TRPML1-dynein signaling pathway. Asah1 gene defect inhibits TRPML1 channel activity and thereby amplifies exosome release, which may contribute to the development of coronary arterial disease under pathological conditions.

2. Materials and methods

2.1 Isolation and culture of CAECs from mouse coronary artery

Isolation of mouse CAECs was carried out and characterized as previously described (Kobayashi, Inoue, Warabi, Minami, & Kodama, 2005; Li, Han, et al., 2013). Briefly, the heart was excised and put into a petri-dish filled with ice-cold Krebs-Henseleit (KH) solution. The cleaned heart with intact aorta was filled with Hank's Balanced Salt Solution (HBSS) using a 25-gauge needle. The opening of the needle was inserted deep into the heart close to the aortic valve. The aorta was tied with the needle as close to the base of the heart as possible. The pump was started with a 20-mL syringe containing warm HBSS at a rate of 0.1 mL/min, and the heart was then flushed for 15 min. HBSS was replaced with warm enzyme solution (1 mg/mL collagenase type I, 0.5 mg/mL soybean trypsin inhibitor, 3% BSA, and 2% antibiotic-antimycotic), which was flushed through the heart at a rate of 0.1 mL/min. Perfusion fluid was collected at 30-, 60-, and 90-min intervals. At 90 min, the heart was cut with scissors, and the apex was opened to flush out the cells that collected inside the ventricle. After centrifugation of the cell suspension at 1000 rpm for 10 min, the cell pellet was resuspended with Dulbecco's modified Eagle's medium and cultured on a 2% gelatin-coated flask at 37 °C before use in experiments.

2.2 GCaMP3 Ca^{2+} imaging

At 18–24h after nucleofection with GCaMP3-ML1, CAECs were used for experiments (Li et al., 2019). The fluorescence intensity at 470nm (F470) was recorded with a digital camera (Nikon Diaphoto TMD Inverted Microscope). Metafluor imaging and analysis software were used to acquire, digitize and store the images for offline processing and statistical analysis (Universal Imaging, Bedford Hills, NY, USA). Lysosomal Ca^{2+} release was measured under a "low" external Ca^{2+} solution, which contained 145mM NaCl, 5mM KCl, 3mM $MgCl_2$, 10mM glucose, 1mM EGTA, and 20mM HEPES (pH 7.4).

2.3 Isolation of lysosomes from CAECs

After treatment with vacuolin-1 (1μM) for 2h, isolation of lysosomes was performed (Schieder, Rotzer, Bruggemann, Biel, & Wahl-Schott, 2010). After washing CAECs with pre-cooled PBS, pre-cooled homogenization buffer (0.25M sucrose, 10mM Tris, pH=7.4) was used for detachment of CAECs by a cell scraper. The cell suspension in glass-grinding vessel was homogenized using a Teflon pestle operated at 900 rotations per minute (rpm). The homogenate was then transferred to a 1.5mL microfuge test tube and centrifuge at 14,000g for 15min at 4°C. After centrifugation, the middle part of supernatant was transferred to a 10-mL polycarbonate centrifuge tube. An equal volume of 16mM $CaCl_2$ was added to precipitate lysosomes. After being shaken on a rotary shaker at 100rpm for 5min at 4°C, the supernatant was centrifuged at 25,000g for 15min at 4°C in an ultracentrifuge. The supernatant was discarded and the pellet was resuspended in one volume of ice cold washing buffer (150mM KCl, 10mM Tris, pH=7.4). The suspension was centrifuged at 25,000g for 15min at 4°C in an ultracentrifuge. After the supernatant was discarded, the pellet containing lysosomes was resuspended in 40μL of washing buffer, which was used for whole-lysosome patch clamp recording. All steps were performed on ice to minimize the activation of intracellular phospholipases and proteases that could damage the lysosomes. Lysosomes were used for electrophysiological recordings within 3h of isolation to keep lysosomes fresh. To confirm the purity of isolated lysosomes, CytoPainter lysosome staining kit (abcam, Cambridge, United Kingdom) was used to stain the lysosomes in CAECs before lysosome isolation. The dye working solution was added to the culture medium and incubated at 37°C for 2h. After lysosome isolation, the green fluorescence at Ex/Em=490/525nm of samples were detected using Guava Easycyte Mini Flow Cytometry System

(Guava Technologies, Hayward, CA, USA). The data were analyzed using Guava acquisition and analysis software (Guava Technologies, Hayward, CA, USA).

2.4 Whole-lysosome patch clamp recording

CAECs were treated with vacuolin-1 (1 μM) for 2 h to increase the size of lysosomes. Using a planar patch clamp system, Port-a-Patch (Nanion Technologies), the whole-lysosome electrophysiology was performed on isolated lysosomes from CAECs (Schieder et al., 2010). The bath solution contained 60 mM KMSA, 60 mM KF, and 10 mM HEPES (pH 7.2, 2 mM CaMSA was added immediately before starting the measurements to avoid precipitation of CaF2). The luminal solution contained 70 mM KMSA, 60 mM CaMSA, 2 mM MgCl2, and 10 mM HEPES (pH 4.6). The planar patch-clamp technology was combined with a pressure control system and microstructured glass chips containing an aperture of around 1 μm diameter (resistances of 10–15 MΩ) (Nanion Technologies). Currents were recorded using an EPC-10 patch-clamp amplifier and PatchMaster acquisition software (HEKA). Data were digitized at 40 kHz and filtered at 2.8 kHz. The membrane potential was held at −60 mV, and 500 ms voltage ramps from −200 to +100 mV were applied every 5 s. All recordings were obtained at room temperature (21–23 °C) and all recordings were analyzed using PatchMaster (HEKA) and Origin 6.1 (OriginLab). The salt-agar bridges were used to connect the reference Ag/AgCl wire to the bath solution to minimize voltage offsets. Ag/AgCl-coated electrodes were chloridated in bleach solution approximately 15 min until a black AgCl-layer was obvious on the silver wire. The liquid junction potential was corrected.

2.5 Structured illumination microscopy

After treatments followed by fixation, the cells were incubated with rabbit anti-Rab7a antibody (1:100; Abcam Biotechnology, Cambridge, United Kingdom) and rat anti-Lamp-1 antibody (1:100; Santa Cruz Biotechnology, Dallas, TX, USA) overnight at 4 °C. After washing the slides, Alexa 488-labeled anti-rabbit secondary antibody (1:200; Life Technologies, CA, USA) and Alexa 594-labeled anti-rat secondary antibody (1:200; Life Technologies, CA, USA) were added to the cell slides and incubated for 1 h at room temperature. Slides were then washed, stained with DAPI, and mounted. A Nikon fluorescence microscope in the structured illumination microscopy (SIM) mode was used to obtain images. Image Pro Plus 6.0 software (Media Cybernetics, Bethesda, MD, USA) was employed to analyze colocalization, expressed as the Pearson correlation coefficient (Li, Huang, Li, Ritter, & Li, 2021).

2.6 Nanoparticle tracking analysis

Nanoparticle tracking analysis measurements were performed with a NanoSight NTA3.2 Dev Build 3.2.16 (Malvern Instruments Ltd., UK), equipped with a sample chamber with a 638-nm laser and a Viton fluoroelastomer O-ring. The samples were injected in the sample chamber with sterile syringes (BD, New Jersey, USA) until the liquid reached the tip of the nozzle. All measurements were performed at room temperature. The screen gain and camera level were 10 and 13, respectively. Each sample was measured at standard measurement, 30 s with manual shutter and gain adjustments. Three measurements of each sample were performed. 3D figures were exported from the software. Particles sized between 50 and 100 nm were calculated (Li et al., 2019).

2.7 Dynamic analysis of lysosome movement in CAECs

CAECs cultured in 35 mm dish were incubated with 50 nM LysoTracker Red DND-99 (Thermo Fisher Scientific, Waltham, MA, USA) for 2 h. The confocal fluorescent microscopic recording was conducted with a confocal laser scanning microscope (Fluoview FV1000, Olympus, Japan). The fluorescent images for lysosomes in CAECs were continuously recorded at an excitation/emission (nm) of 555/565 by using XYT recording mode with a speed of 1 frame/10 s for 10 min. Lysosome tracking was performed in Image J using manual tracking plugin as described in previous studies (Jahreiss, Menzies, & Rubinsztein, 2008; Xu et al., 2013). Ten lysosomes were chosen at random for each cell. These lysosomes were then tracked manually, while the program calculated velocity of lysosome trafficking for each frame.

2.8 Statistical analysis

All of the values are expressed as mean ± SEM. Significant differences among multiple groups were examined using *t*-test or ANOVA followed by a Student-Newman-Keuls test. $P < 0.05$ was considered statistically significant.

3. Results

3.1 Elevation of exosome release and reduction of lysosome-MVB interaction in CAECs lacking Asah1 gene

We isolated and cultured CAECs from WT/WT and endothelial cell-specific Asah1 gene (gene encoding AC) knockout (Asah1$^{fl/fl}$/EC^{cre}) mice. By nanoparticle tracking analysis, we first measured exosome release from CAECs with or without Asah1 gene. In Fig. 1A, the representative 3-D histograms showed that the exosome concentration in culture medium of WT/WT CAECs was lower than CAECs lacking Asah1 gene. The summarized data suggest that Asah1 gene deletion remarkably increased exosome release from CAECs. Moreover, we observed the interaction of lysosomes

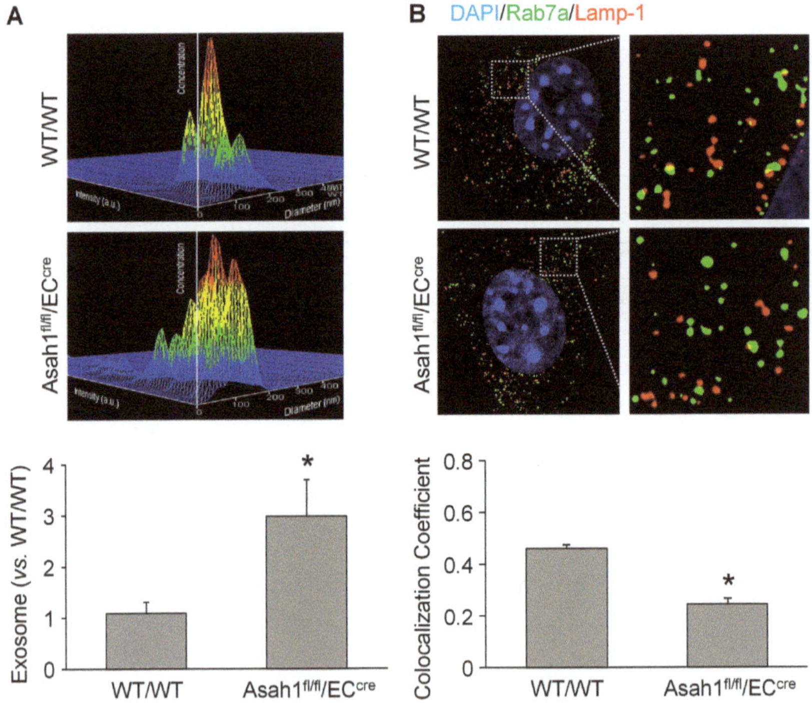

Fig. 1 Elevation of exosome release and reduction of lysosome-MVB interaction in CAECs lacking Asah1 gene. (A) Representative images and summarized data showing that exosome release from CAECs was remarkably elevated in by Asah1 gene deletion ($n=6$). The x axis is diameter (nm); the y axis is concentration; the z axis is intensity (a.u.). (B) Representative images and summarized data showing that lysosome-MVB interaction in CAECs was significantly attenuated by Asah1 gene deletion ($n=3$). * $p<0.05$ vs. WT/WT group.

(Lamp-1, a lysosome marker stained with red fluorescence) and MVBs (Rab7a, a MVB marker stained with green fluorescence) in CAECs with or without AC function by super-resolution microscopy. As shown in Fig. 1B, considerable amount of lysosome-MVB interaction was observed in the cytosol of WT/WT CAECs. Asah1 gene knockout, however, decreased the encounter and fusion of lysosomes and MVBs in CAECs. The summarized data demonstrated that colocalization of Rab7a and Lamp-1 was significantly reduced by Asah1 gene knockout in CAECs.

3.2 Inhibition of exosome release from CAECs lacking Asah1 gene by sphingosine

Given that TRPML1 channel governs lysosomal Ca^{2+} release to regulate lysosomal functions such as trafficking (Li, Huang, et al., 2020; Li et al., 2019), we hypothesized that AC may control TRPML1 channel activity to regulate lysosome trafficking and fusion to MVBs and thereby determine exosome release from CAECs. To test this hypothesis, we treated CAECs of WT/WT and $Asah1^{fl/fl}/EC^{cre}$ mice with vehicle, ML-SA1, or sphingosine for 24h before nanoparticle tracking analysis. As shown in Fig. 2, Asah1 gene knockout remarkably amplified exosome release from CAECs. ML-SA1, a TRPML1 channel agonist, failed to alter the elevation of exosome secretion from CAECs lacking Asah1 gene. Sphingosine as the product of ceramide metabolism by AC, however, substantially reduced exosomes released into culture medium by CAECs lacking Asah1 gene. In WT/WT CAECs, ML-SA1 and sphingosine had no significant effects on exosome release.

3.3 Regulation of lysosome trafficking and lysosome-MVB interaction by ML-SA1 and sphingosine in CAECs

To examine the role of TRPML1 channel in the regulation of lysosome function and exosome release in CAECs, we tested whether lysosome trafficking is controlled by TRPML1 channel in these cells. LysoTracker Red DND-99 was loaded into live CAECs and the lysosome movement was continuously monitored for 10min under different conditions. After treatment with ML-SA1, lysosome movement was obviously faster in WT/WT CAECs compared to CAECs lacking Asah1 gene (Fig. 3A). The summarized data suggested that Asah1 gene deletion significantly inhibited lysosome trafficking in CAECs. ML-SA1 enhanced lysosome trafficking in WT/WT CAECs but failed to affect lysosome movement in CAECs lacking Asah1 gene. However, sphingosine enhanced lysosome trafficking in CAECs of both WT/WT and $Asah1^{fl/fl}/EC^{cre}$ mice (Fig. 3B). By super-resolution microscopy, we found that ML-SA1 significantly increased

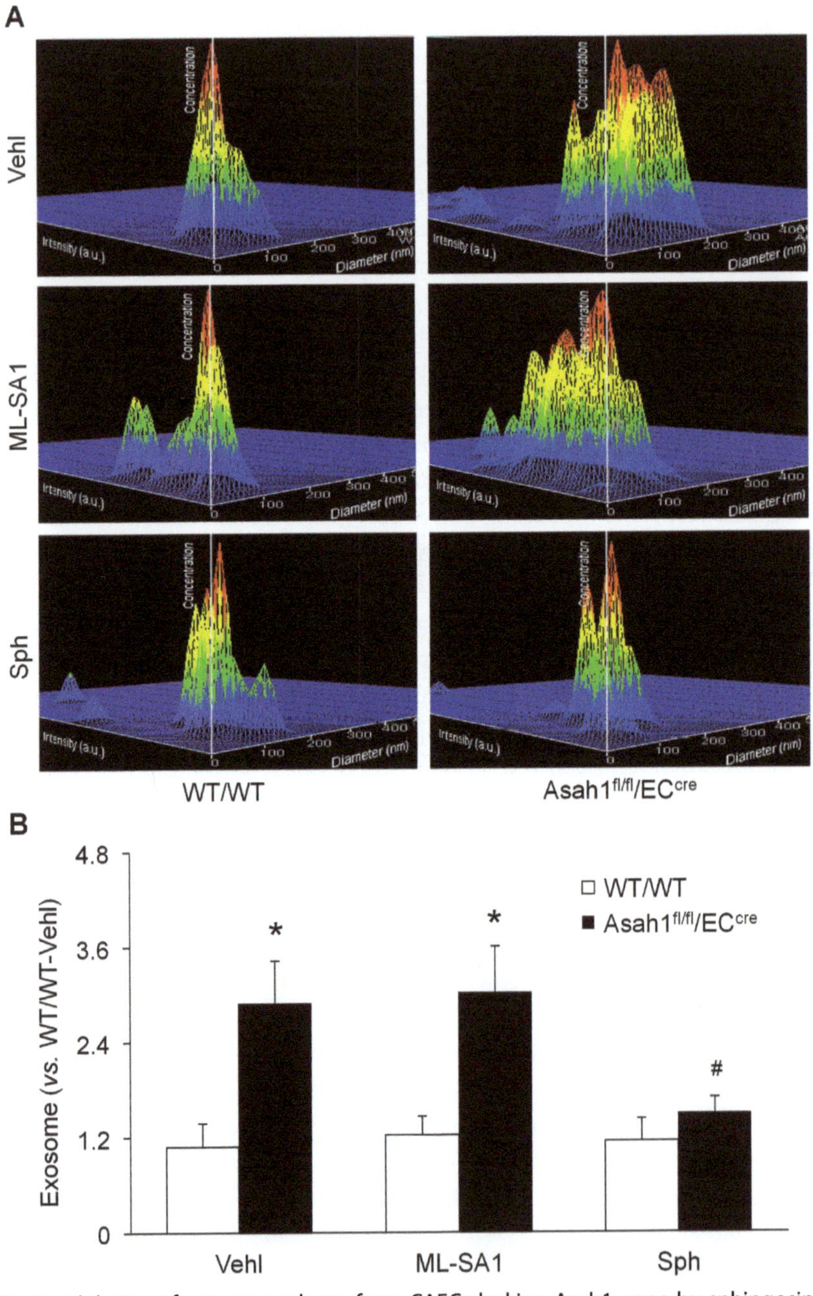

Fig. 2 Inhibition of exosome release from CAECs lacking Asah1 gene by sphingosine. (A) Representative images showing exosomes released into the culture medium of CAECs of WT/WT and Asah1$^{fl/fl}$/EC^{cre} mice under different conditions. The x axis is diameter (nm); the y axis is concentration; the z axis is intensity (a.u.). (B) Summarized data showing that sphingosine (10 μM), but not ML-SA1 (10 μM), inhibited exosome secretion from CAECs lacking Asah1 gene (n = 6). * $p < 0.05$ vs. WT/WT group. # $p < 0.05$ vs. Vehl group. Vehl, vehicle; Sph, sphingosine.

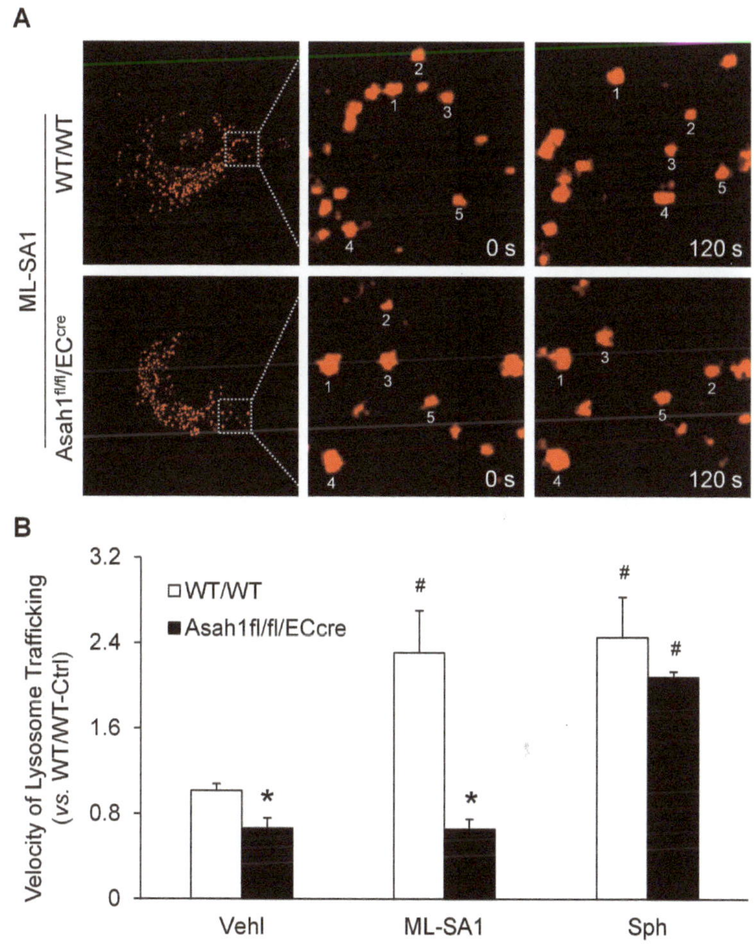

Fig. 3 Effects of ML-SA1 and sphingosine on lysosome trafficking in CAECs. (A) Representative images showing that ML-SA1 enhanced lysosome trafficking in WT/WT CAECs, but not CAECs lacking Asah1 gene. (B) Summarized data showing velocity of lysosome trafficking in CAECs of WT/WT and Asah1$^{fl/fl}$/EC^{cre} mice under different conditions (n=3–4). * $P<0.05$ vs. WT/WT group. # $P<0.05$ vs. Vehl group.

lysosome-MVB interaction in WT/WT CAECs, but not in CAECs lacking Asah1 gene. Nevertheless, sphingosine enhanced lysosome-MVB interaction in CAECs regardless of Asah1 gene expression (Fig. 4).

3.4 Characterization of TRPML channels in CAECs

To further address the critical role of lysosomal TRPML1 in AC-mediated regulation of lysosome trafficking and exosome secretion, we characterized this channel in mouse CAECs. By RT-PCR, we found that the expression

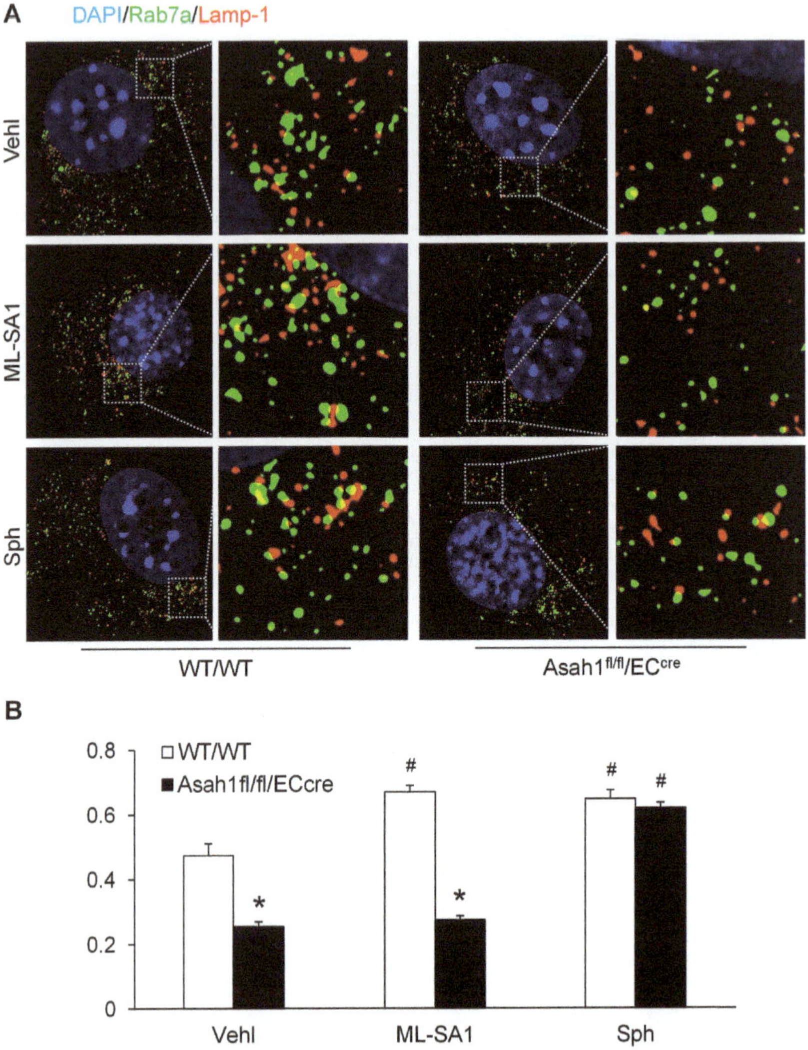

Fig. 4 Regulation of lysosome-MVB interaction by ML-SA1 and sphingosine in CAECs. (A) Representative images showing lysosome-MVB interaction in CAECs of WT/WT and Asah1$^{fl/fl}$/EC^{cre} mice under different conditions. (B) Summarized data showing lysosome-MVB interaction in CAECs of WT/WT and Asah1$^{fl/fl}$/EC^{cre} mice under different conditions (n = 3). * $P < 0.05$ vs. WT/WT group. # $P < 0.05$ vs. Vehl group.

level of TRPML1 was remarkably higher than TRPML2 and TRPML3 in cultured CAECs (Fig. 5A). To further confirm our findings, confocal microscopy demonstrated that TRPML1 was abundantly expressed in CAECs and highly colocalized with Lamp-1, a lysosome marker. On the contrary,

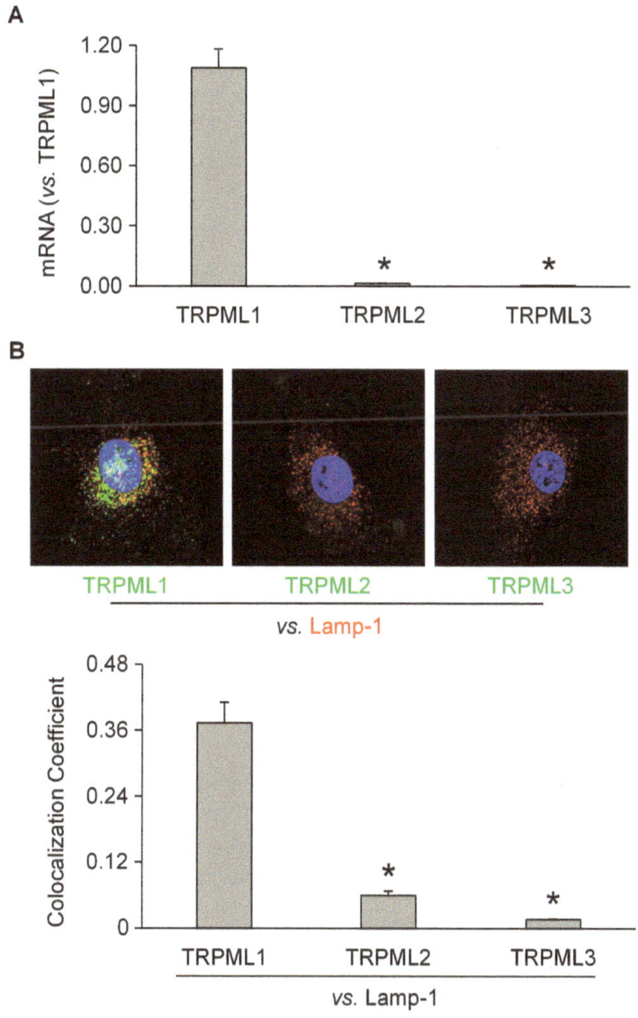

Fig. 5 Characterization of TRPML channels in CAECs. (A) Summarized data showing relative mRNA levels of TRPML1, TRPML2, and TRPML3 in murine CAECs detected by RT-PCR (n=6). (B) Representative images and summarized data showing that TRPML1, but not TRPML2 or TRPML3, was abundantly expressed in murine CAECs ($n=4$). * $p<0.05$ vs. relative mRNA level of TRPML1 (Fig. 2A) or colocalization coefficient of TRPML1 and Lamp-1 (Fig. 2B).

the expressions of TRPML2 and TRPML3 and their colocalizations with Lamp-1 were markedly lower than TRPML1 (Fig. 5B). These experiments provided the first evidence showing the expression of TRPML1 in murine CAECs.

3.5 Blockade of TRPML1 channel by Asah1 gene deletion in CAECs

To specifically detect lysosomal Ca^{2+} release through TRPML1 channel, nucleofection of GCaMP3-ML1 plasmid into CAECs was performed to express GCaMP3, a single-wavelength genetically encoded Ca^{2+} indicator, on the cytoplasmic amino terminus of TRPML1 in these cells as described in our previous studies (Li, Huang, et al., 2020; Li et al., 2019). A fluorescent microscopic imaging system was used to dynamically and continuously monitor the GCaMP3 fluorescence signal (F470) in CAECs. The intensity of Ca^{2+}-induced GCaMP3 fluorescence indicated the amount of Ca^{2+} released through lysosomal TRPML1 channel. As shown in Fig. 6A, the GCaMP3 fluorescence detected in WT/WT CAECs was dim under control condition. ML-SA1 remarkably increased the GCaMP3 fluorescence in WT/WT CAECs, but not in CAECs lacking Asah1 gene. Fig. 6B showed that addition of ML-SA1 to the extracellular solution induced a rapid elevation of GCaMP3 fluorescence in WT/WT CAECs, which was followed by a large signal increase caused by late addition of ionomycin, a Ca^{2+} ionophore. In contrast, ML-SA1 failed to augment GCaMP3 fluorescence in CAECs lacking Asah1 gene, while ionomycin still induced a dramatic elevation of GCaMP3 fluorescence in these cells. The summarized data suggested that Asah1 gene deletion impaired lysosomal Ca^{2+} release through TRPML1 channel in CAECs (Fig. 6C). To further confirm our findings, we isolated lysosomes from CAECs of WT/WT and $Asah1^{fl/fl}/EC^{cre}$ mice for whole-lysosome patch clamping. After treatment with vacuolin-1, lysosomes were enlarged and then isolated from CAECs for experiments as described in our previous studies (Li, Huang, et al., 2020; Li et al., 2019). It was found that bath application of ML-SA1 induced whole-lysosome TRPML1 channel-mediated Ca^{2+} currents of WT/WT CAECs in a dose-dependent manner (Fig. 6D). However, there was no significant elevation observed in lysosomal Ca^{2+} currents after addition of ML-SA1 into bath solution of lysosomes isolated from CAECs lacking Asah1 gene (Fig. 6E).

3.6 Rescue of TRPML1 channel activity by sphingosine in CAECs lacking Asah1 gene

Next, we examined the effect of sphingosine on TRPML1 channel activity. By GCaMP3 Ca^{2+} imaging, sphingosine was found to induce lysosomal Ca^{2+} release through TRPML1 channel in CAECs of both WT/WT and $Asah1^{fl/fl}/EC^{cre}$ mice, indicating that sphingosine rescued TRPML1 channel activity in CAECs lacking the Asah1 gene (Fig. 7A–C).

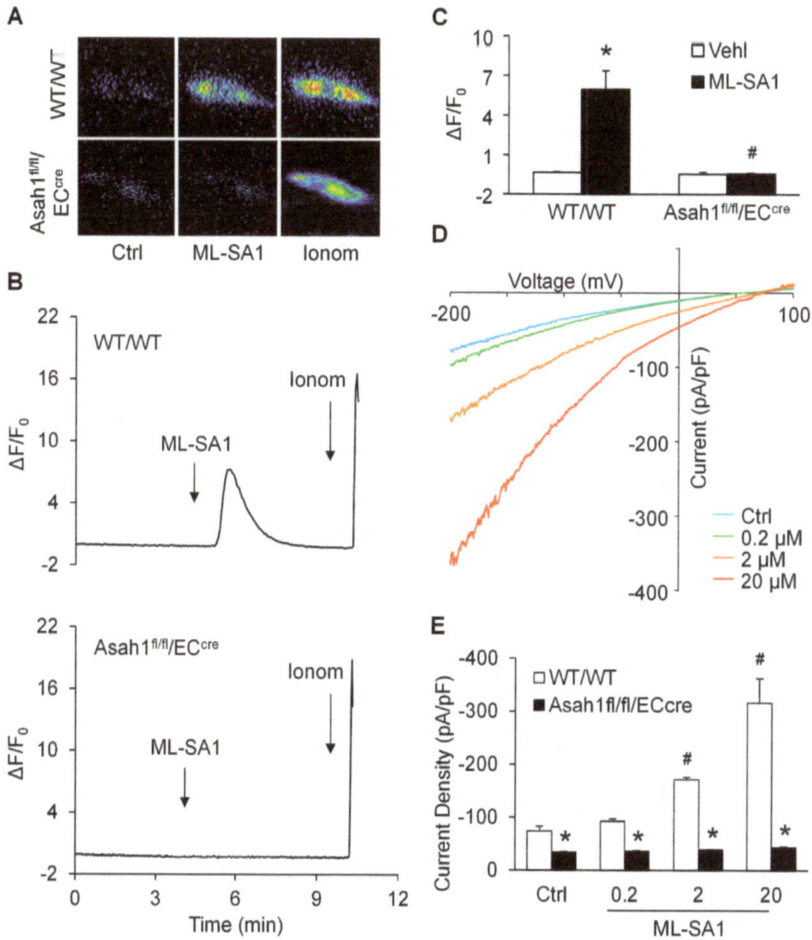

Fig. 6 Blockade of TRPML1 channel by Asah1 gene deletion in CAECs. (A) Representative images showing that ML-SA1 (20 μM) induced TRPML1 channel-mediated Ca^{2+} release in WT/WT CAECs, but not CAECs lacking Asah1 gene. (B) Representative curves showing that ML-SA1 induced elevation of GCaMP3 signal in WT/WT CAECs, but not in CAECs lacking Asah1 gene. (C) Summarized data showing that Asah1 gene deletion blocked TRPML1-mediated Ca^{2+} release induced by ML-SA1 in CAECs (n=4–5). (D) Representative whole lysosome currents of WT/WT CAECs enhanced by ML-SA1 in different doses. (E) ML-SA1 enhanced TRPML1 channel activity of lysosomes isolated from WT/WT CAECs in a concentration-dependent manner. However, ML-SA1 failed to enhance TRPML1 channel activity of lysosomes isolated from CAECs lacking Asah1 gene (n=3). * $p<0.05$ vs. Vehl group (Fig. 3D) or WT/WT group (Fig. 3F), # $p<0.05$ vs. WT/WT group (Fig. 3D) or Ctrl group (Fig. 3F). Ctrl, control; Ionom, ionomycin; Vehl, vehicle.

To further dissect the actions of AC-associated sphingolipids on the opening of TRPML1 channel, whole-lysosome patch clamping was performed to measure ML-SA1-induced whole-lysosome TRPML1 channel-mediated Ca^{2+} currents of CAECs with or without Asah1 gene in the presence of

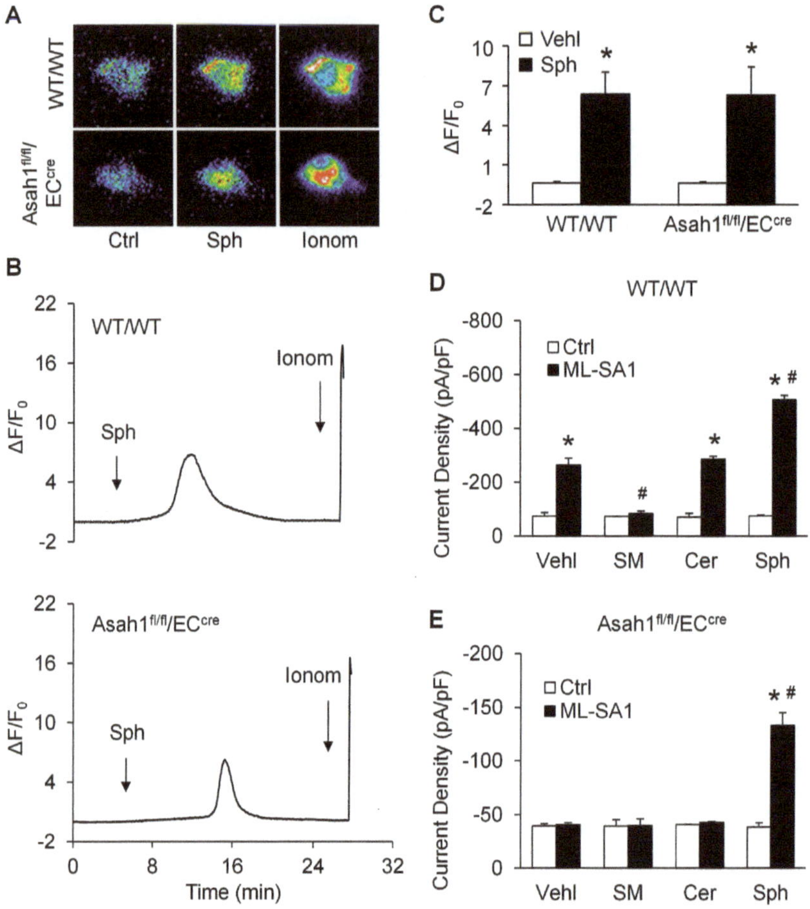

Fig. 7 Rescue of TRPML1 channel activity by sphingosine in CAECs lacking Asah1 gene. (A) Representative images showing that sphingosine (20 μM) induced TRPML1 channel-mediated Ca^{2+} release in CAECs regardless of Asah1 gene expression. (B) Representative curves showing that sphingosine induced elevation of GCaMP3 signal in CAECs regardless of Asah1 gene expression. (C) Summarized data showing that sphingosine induced TRPML1-mediated Ca^{2+} release in CAECs regardless of Asah1 gene expression (n = 3–4). (D) Summarized data of the effects of various sphingolipids on whole lysosome currents of lysosomes isolated from WT/WT CAECs (n = 4). (E) Summarized data of the effects of various sphingolipids on whole lysosome currents of lysosomes isolated from CAECs lacking Asah1 gene (n = 4). * $p < 0.05$ vs. Vehl group (Fig. 4D) or WT/WT group (Fig. 4E and F), # $p < 0.05$ vs. Ctrl group (Fig. 4E and F). Ctrl, control; Ionom, ionomycin; Vehl, vehicle; Sph, sphingosine; Cer, ceramide; SM, sphingomyelin.

AC-associated sphingolipids to determine their regulatory roles on TRPML1 channel activity. We demonstrated that sphingomyelin, ceramide, and sphingosine had different effects on whole-lysosome TRPML1 channel-mediated Ca^{2+} currents of WT/WT CAECs, with inhibition by sphingomyelin, no effects by ceramide, but enhancement by sphingosine. Although both sphingomyelin and ceramide had no effects on whole-lysosome TRPML1 channel-mediated Ca^{2+} currents of CAECs lacking Asah1 gene, sphingosine remained to increase TRPML1 channel activity in these lysosomes (Fig. 7D and E).

3.7 Contribution of dynein activity to lysosomal regulation of exosome release in CAECs

We also tested whether dynein plays an important role in the lysosomal regulation of exosome release in CAECs. Super-resolution microscopy showed that EHNA, a dynein inhibitor, significantly reduced lysosome-MVB interaction in CAECs. Moreover, the elevation of lysosome-MVB interaction by ML-SA1 was substantially inhibited by EHNA (Fig. 8A). Correspondingly, we found that inhibition of dynein activity by EHNA remarkably elevated exosome release from CAECs (Fig. 8B).

4. Discussion

The major goals of the present study were to determine whether lysosomal AC deficiency enhances exosome release through inhibition of lysosome-dependent MVB degradation in CAECs and to explore the mechanisms by which AC regulates lysosome trafficking, lysosome-MVB interaction, and exosome secretion in these cells. It was found that Asah1 gene deletion inhibited lysosome-MVB interaction in CAECs, which was accompanied by enhancement of exosome release from these cells. We further demonstrated that lysosomal TRPML1 channel was blocked in CAECs lacking Asah1 gene. Interestingly, sphingosine as the product of ceramide metabolism by AC rescued TRPML1 channel activity in CAECs with AC deficiency. Moreover, we found that AC-dependent TRPML1 channel activity importantly contributed to lysosome trafficking and lysosomal regulation of exosome release in CAECs. Inhibition of dynein activity also affected lysosome-MVB interaction and exosome secretion in these cells, which confirmed that lysosomal regulation of exosome secretion was attributed to lysosome trafficking (Fig. 9).

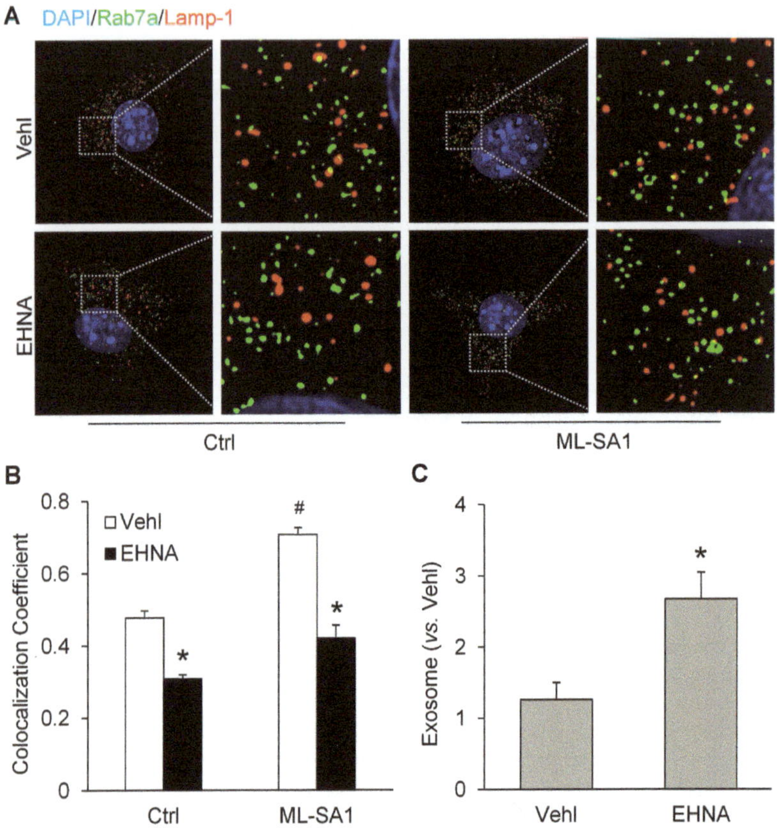

Fig. 8 Contribution of dynein activity to lysosomal regulation of exosome release in CAECs. (A) Representative images showing lysosome-MVB interaction in CAECs under different conditions. (B) Summarized data showing lysosome-MVB interaction in CAECs under different conditions (n = 3). (C) Summarized data showing that EHNA (30 μM) increased exosome release from CAECs ($n = 5$). * $p < 0.05$ vs. Vehl group, # $p < 0.05$ vs. Ctrl group. Vehl, vehicle; Ctrl, control; EHNA, erythro-9-(2-hydroxy-3-nonyl)adenine.

As a lysosomal enzyme, AC determines tissue or cellular ceramide levels under physiological and pathological conditions (Coant, Sakamoto, Mao, & Hannun, 2017; Li, Kidd et al., 2020). It has been reported that mutations in the AC gene (ASAH1) or deficiency of lysosomal AC activity (metabolism of ceramide to sphingosine) in human cells are a major genetic or pathogenic mechanism for the development of Farber disease, which results in derangement of metabolic homeostasis (Masuko et al., 2009; Schuchman, 2016), EC dysfunction, and vascular injury (Feng et al., 2019; Sikora et al., 2017; Yuan et al., 2019). Recently, we have demonstrated that AC

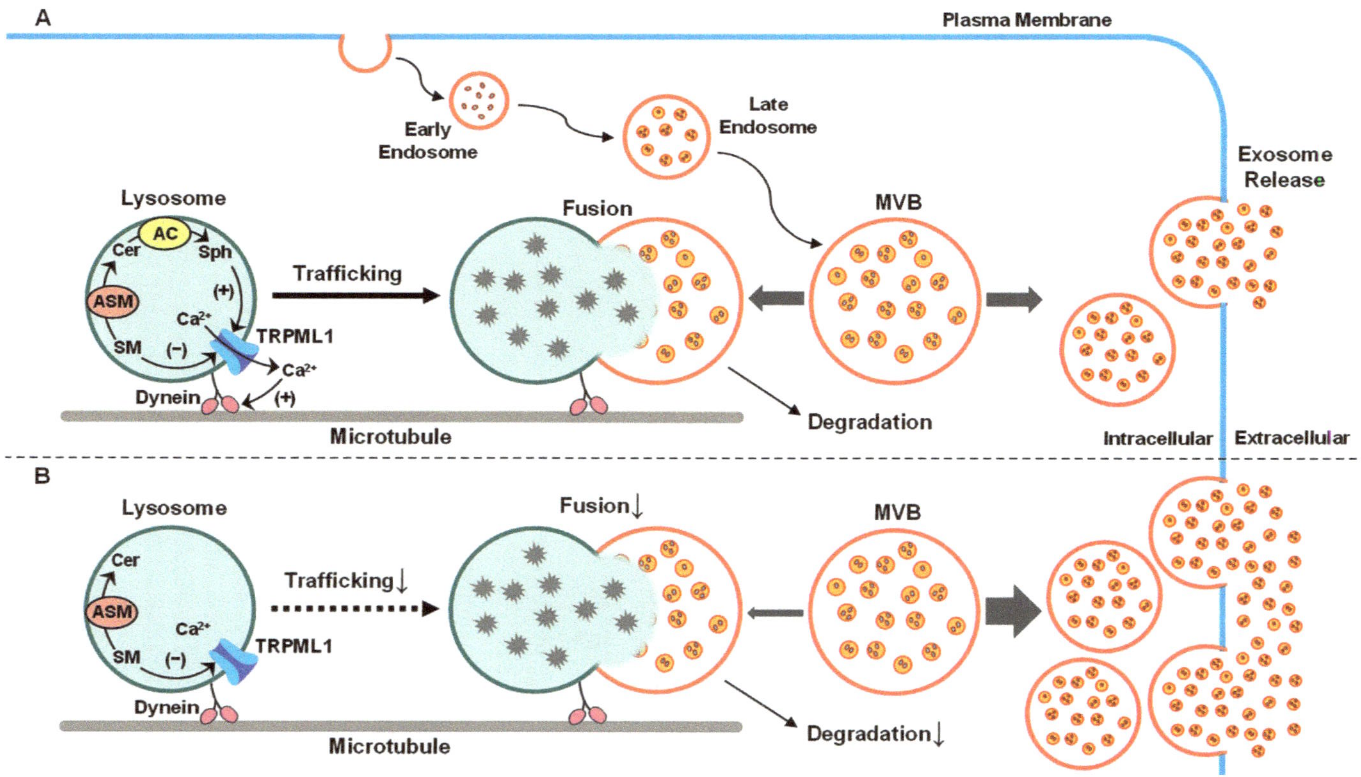

Fig. 9 See figure legend on next page.

controls exosome secretion from CAECs and thereby regulates exosome-mediated release of NLRP3 inflammasome products from these cells during hyperglycemia (Yuan et al., 2019). EC-specific Asah1 gene deletion aggravates hyperglycemia-induced arterial inflammatory response through amplification of inflammatory exosome release from CAECs (Yuan et al., 2019). However, the molecular mechanism underlying the regulation of exosome release by AC remains poorly understood. In the present study, we demonstrated that decreased lysosome-MVB interaction was associated with elevated exosome release in CAECs lacking Asah1 gene. Our findings indicate that the lysosome determines MVB fate and controls exosome release in CAECs. In this regard, it has been reported that inhibition of lysosome function with different alkaline agents or lysosomal v-ATPase inhibitor, bafilomycin A, or chloroquine can increase exosome secretion in different cells such as neurons, epithelial cells, and vascular cells (Alvarez-Erviti et al., 2011; Vingtdeux et al., 2007; Wei et al., 2013). Mutations of VPS4 gene required for MVB maturation and fusion with lysosome increased the levels of EV-associated proteins within cells (Hasegawa et al., 2011). In addition, there is evidence that MVBs may be fused with autophagosomes to form amphisomes and subsequently fuse with lysosomes to terminate MVB fate (Baixauli et al., 2014; Fader et al., 2008; Murrow & Debnath, 2015). Although our findings suggest that AC may play a vital role in the lysosomal regulation of exosome secretion in CAECs, it remains unknown how AC regulates lysosome function and thereby determines lysosome-MVB interaction and exosome release in these cells.

It has been reported that MVBs and lysosomes are concentrated near the microtubule-organizing center in the juxtanuclear region of the cell, suggesting that both MVBs and lysosomes need to traffic along microtubules

Fig. 9 Regulation of exosome release by AC-TRPML1-dynein signaling pathway in CAECs. (A) Exosomes in CAECs are formed as intraluminal vesicles of MVBs via endocytic process. MVBs can fuse with and deliver content to lysosomes for degradation or fuse with plasma membrane to release intraluminal vesicles as exosomes, which is regulated by lysosome trafficking and fusion to MVBs, a mechanism constitutively controlling the fate of MVBs. The AC-associated sphingolipids such as sphingosine and sphingomyelin gate TRPML1 channel activity, which drives lysosome trafficking to and fusion with MVBs for degradation preventing exosome secretion in a dynein-dependent manner. (B) If AC in CAECs is defective or deficient, lysosome trafficking, lysosome-MVB interaction, and lysosome-dependent MVB degradation may be inhibited, leading to enhancement of exosome release. MVB, multivesicular body; Sph, sphingosine; Cer, ceramide; SM, sphingomyelin; TRPML1, transient receptor potential-mucolipin 1.

to encounter, kiss and fuse (Bright, Reaves, Mullock, & Luzio, 1997; Futter et al., 1996; Mullock, Bright, Fearon, Gray, & Luzio, 1998; Ward, Pevsner, Scullion, Vaughn, & Kaplan, 2000). In fact, there is considerable evidence that lysosomes have trafficking function within different cells (Bhat, Li, et al., 2020; Bhat, Yuan, Cain, et al., 2020; Bhat, Yuan, Camus, et al., 2020; Yuan et al., 2019). In our previous studies, we found that lysosome trafficking is a main regulatory mechanism of autophagic flux and exosome secretion in vascular smooth muscle cells and renal podocytes, which may depend upon the Ca^{2+} bursts from lysosomes (Bhat, Li, et al., 2020; Li et al., 2019). It has been shown that Ca^{2+} enters lysosomal compartment by H^+/Ca^{2+} exchange under resting condition and is released through TRPML1 channels in response to endogenously produced NAADP (Churchill et al., 2002; Dell'Angelica, Mullins, Caplan, & Bonifacino, 2000; Galione, 2006; Zhang et al., 2006; Zhang, Xia, & Li, 2010) or other factors like PIPs (PI (3,5)P2) and irons (Dong et al., 2008; Dong et al., 2010; Li et al., 2016). Based on these previous studies, we hypothesized that AC controls lysosome trafficking and fusion to MVBs via regulation of lysosomal Ca^{2+} release through TRPML1 channel. We first demonstrated that TRPML1, but not TRPML2 or TRPML3, was abundant in CAECs. Then, we found that ML-SA1 as a TRPML1 channel agonist activated lysosomal Ca^{2+} release through TRPML1 channel in CAECs. Moreover, lysosome trafficking and lysosome-MVB interactions were enhanced by ML-SA1 in these cells. However, Asah1 gene deletion blocked these actions of ML-SA1 on CAECs. To our knowledge, these results provide the first evidence that regulation of TRPML1 channel-mediated Ca^{2+} release by AC may serve as a critical mechanism determining exosome release from ECs. In some previous studies, the Ca^{2+} released from lysosomes has been found to drive lysosome moving to meet with other cellular vesicles such as endosomes, MVBs, autophagosomes, and sarcoplasmic reticulum (Churchill et al., 2002; Kinnear et al., 2008; Zhang et al., 2006, 2010; Zhang, Jin, Yi, & Li, 2009; Zhang & Li, 2007). In other cells such as breast cancer cells, epithelial cells and neurons, the Ca^{2+}-dependent lysosome trafficking has also been observed (Dong et al., 2010; Glunde et al., 2003; Li et al., 2016; Shen et al., 2012; Trajkovic et al., 2006). In addition to TRPML1 channel, other channels such as two-pore channels may also be implicated in lysosomal Ca^{2+} release, but they may form hybrids with TRPML1 and need accessory proteins to exert their action (Guse, 2012; Li, Zhang, Abais, Ritter, & Zhang, 2013).

After confirmation of deficient TRPML1 channel activity as an important mechanism, we went on to address how AC controlled TRPML1 channel-mediated Ca^{2+} release and how Asah1 gene deletion blocked TRPML1 channel in CAECs. The lysosomal TRPML1 channel activity has been shown to be regulated by sphingolipids. Accumulated sphingomyelin was found to inhibit lysosomal TRPML1 channel activity and reduce lysosomal Ca^{2+} release, blunting lysosome trafficking that leads to lysosomal storage disease as shown in Niemann-Pick disease and other diseases, including minimal change disease in the kidney (Li et al., 2019; Piccoli et al., 2011; Shen et al., 2012). Sphingosine, the product of ceramide metabolism by AC, enhanced lysosomal Ca^{2+} release through TRPML1 channel and promoted lysosome trafficking in some epithelial cells (Li et al., 2019; Shen et al., 2012). We tested whether deficiency of TRPML1 channel activity was attributed to altered ceramide metabolism by AC. We found that sphingosine activated TRPML1 channel-mediated Ca^{2+} release in CAECs with or without Asah1 gene, suggesting that sphingosine production by AC is essential for the normal function of TRPML1 channel. Correspondingly, sphingosine recovered lysosome trafficking and lysosome-MVB interaction in CAECs lacking Asah1 gene, leading to reduction of exosome release from these cells. Moreover, we found that TRPML1 channel activity was inhibited by sphingomyelin but not affected by ceramide, the substrate of AC. To our knowledge, there have been no reports on the role of sphingolipids in the regulation of TRPML1 channel activity in ECs. Our results provide direct evidence that AC-dependent production of sphingosine critically contributes to TRPML1 channel activation and consequent lysosomal regulation of exosome secretion in ECs. In studies using other cell types, it has been shown that the spontaneous elevation of intracellular sphingosine levels via caged sphingosine leads to a significant and transient calcium release from lysosomes, which is independent of sphingosine-1-phosphate, extracellular calcium level, and ER calcium level (Hoglinger et al., 2015). Moreover, we have demonstrated that Asah1 gene deletion induces ceramide accumulation and sphingosine reduction in podocytes (Li, Huang, et al., 2020). Our findings together with these previous results provide strong evidence that reduced production of sphingosine due to AC deficiency leads to inhibition of TRPML1 channel-mediated Ca^{2+} release, thereby enhancing exosome release in CAECs due to deficient lysosome trafficking and attenuated lysosome-MVB interaction.

Another interesting finding of the present study is that the enhancement of lysosome-MVB interaction by ML-SA1 was prevented by inhibition of dynein activity in CAECs. Also, elevation of exosome release was associated with inhibition of dynein activity in these cells. As a motor protein, dynein mediates the movement of organelles, including lysosomes, from the plus end to the minus end of microtubules, which is known as retrograde transport (Pu, Guardia, Keren-Kaplan, & Bonifacino, 2016). Interaction with dynactin and formation of dynein-dynactin complex are required for the association of dynein and lysosome (Burkhardt, Echeverri, Nilsson, & Vallee, 1997; Lin & Collins, 1992). More recently, it has been reported that ALG-2, a lysosomal Ca^{2+}-sensor, promotes interaction of TRPML1 channel with dynein–dynactin complex to regulate centripetal transport of lysosomes in response to lysosomal Ca^{2+} release through the TRPML1 channel (Li et al., 2016). Based on these previous studies, our findings further confirm that lysosome trafficking is essential for the lysosomal regulation of exosome release in ECs, which depends on the normal activity of AC, TRPML1 channel, and dynein. Activation of TRPML1 channels contributes to the lysosomal regulation of exosome secretion through dynein-ependent lysosome trafficking and fusion to MVB in these cells. The lysosome-MVB fusion may lead to MVB degradation and thereby maintain the speed of exosome release from ECs under control.

In summary, the present study revealed a new regulatory mechanism of exosome release from ECs. The functional deficiency of AC and consequent dysregulation of AC-TRPML1-dynein signaling pathway in CAECs may represent a novel early event in the pathogenesis of coronary arterial disease. These results provide the impetus for development of new therapeutic strategies targeting AC activity for prevention or treatment of coronary arterial disease associated with lysosome dysfunction and elevation of exosome release in CAECs under different pathological conditions, such as hypercholesterolemia, hyperhomocysteinemia, and hyperglycemia.

References

Alvarez-Erviti, L., Seow, Y., Schapira, A. H., Gardiner, C., Sargent, I. L., Wood, M. J., et al. (2011). Lysosomal dysfunction increases exosome-mediated alpha-synuclein release and transmission. *Neurobiology of Disease*, *42*(3), 360–367. https://doi.org/10.1016/j.nbd.2011.01.029.

Baixauli, F., Lopez-Otin, C., & Mittelbrunn, M. (2014). Exosomes and autophagy: Coordinated mechanisms for the maintenance of cellular fitness. *Frontiers in Immunology*, *5*, 403. https://doi.org/10.3389/fimmu.2014.00403.

Bhat, O. M., Li, G., Yuan, X., Huang, D., Gulbins, E., Kukreja, R. C., et al. (2020). Arterial medial calcification through enhanced small extracellular vesicle release in smooth muscle-specific Asah1 gene knockout mice. *Scientific Reports*, *10*(1), 1645. https://doi.org/10.1038/s41598-020-58568-5.

Bhat, O. M., Yuan, X., Cain, C., Salloum, F. N., & Li, P. L. (2020). Medial calcification in the arterial wall of smooth muscle cell-specific Smpd1 transgenic mice: A ceramide-mediated vasculopathy. *Journal of Cellular and Molecular Medicine, 24*(1), 539–553. https://doi.org/10.1111/jcmm.14761.

Bhat, O. M., Yuan, X., Camus, S., Salloum, F. N., & Li, P. L. (2020). Abnormal lysosomal positioning and small extracellular vesicle secretion in arterial stiffening and calcification of mice lacking Mucolipin 1 gene. *International Journal of Molecular Sciences, 21*(5), 1713. https://doi.org/10.3390/ijms21051713.

Boulanger, C. M., Loyer, X., Rautou, P. E., & Amabile, N. (2017). Extracellular vesicles in coronary artery disease. *Nature Reviews. Cardiology, 14*(5), 259–272. https://doi.org/10.1038/nrcardio.2017.7.

Bright, N. A., Reaves, B. J., Mullock, B. M., & Luzio, J. P. (1997). Dense core lysosomes can fuse with late endosomes and are re-formed from the resultant hybrid organelles. *Journal of Cell Science, 110*(Pt 17), 2027–2040.

Burkhardt, J. K., Echeverri, C. J., Nilsson, T., & Vallee, R. B. (1997). Overexpression of the dynamitin (p50) subunit of the dynactin complex disrupts dynein-dependent maintenance of membrane organelle distribution. *The Journal of Cell Biology, 139*(2), 469–484. https://doi.org/10.1083/jcb.139.2.469.

Chistiakov, D. A., Orekhov, A. N., & Bobryshev, Y. V. (2015). Extracellular vesicles and atherosclerotic disease. *Cellular and Molecular Life Sciences, 72*(14), 2697–2708. https://doi.org/10.1007/s00018-015-1906-2.

Churchill, G. C., Okada, Y., Thomas, J. M., Genazzani, A. A., Patel, S., & Galione, A. (2002). NAADP mobilizes Ca(2+) from reserve granules, lysosome-related organelles, in sea urchin eggs. *Cell, 111*(5), 703–708. https://doi.org/10.1016/s0092-8674(02)01082-6.

Coant, N., Sakamoto, W., Mao, C., & Hannun, Y. A. (2017). Ceramidases, roles in sphingolipid metabolism and in health and disease. *Advances in Biological Regulation, 63*, 122–131. https://doi.org/10.1016/j.jbior.2016.10.002.

Crescitelli, R., Lasser, C., Szabo, T. G., Kittel, A., Eldh, M., Dianzani, I., et al. (2013). Distinct RNA profiles in subpopulations of extracellular vesicles: Apoptotic bodies, microvesicles and exosomes. *Journal of Extracellular Vesicles, 2*(1), 20677. https://doi.org/10.3402/jev.v2i0.20677.

Davenport, A. P., Hyndman, K. A., Dhaun, N., Southan, C., Kohan, D. E., Pollock, J. S., et al. (2016). Endothelin. *Pharmacological Reviews, 68*(2), 357–418. https://doi.org/10.1124/pr.115.011833.

Dell'Angelica, E. C., Mullins, C., Caplan, S., & Bonifacino, J. S. (2000). Lysosome-related organelles. *The FASEB Journal, 14*(10), 1265–1278. https://doi.org/10.1096/fj.14.10.1265.

Deng, H., Sun, C., Sun, Y., Li, H., Yang, L., Wu, D., et al. (2018). Lipid, protein, and MicroRNA composition within mesenchymal stem cell-derived exosomes. *Cellular Reprogramming, 20*(3), 178–186. https://doi.org/10.1089/cell.2017.0047.

Dong, X. P., Cheng, X., Mills, E., Delling, M., Wang, F., Kurz, T., et al. (2008). The type IV mucolipidosis-associated protein TRPML1 is an endolysosomal iron release channel. *Nature, 455*(7215), 992–996. https://doi.org/10.1038/nature07311.

Dong, X. P., Shen, D., Wang, X., Dawson, T., Li, X., Zhang, Q., et al. (2010). PI(3,5)P(2) controls membrane trafficking by direct activation of mucolipin ca(2+) release channels in the endolysosome. *Nature Communications, 1*, 38. https://doi.org/10.1038/ncomms1037.

Fader, C. M., Sanchez, D., Furlan, M., & Colombo, M. I. (2008). Induction of autophagy promotes fusion of multivesicular bodies with autophagic vacuoles in k562 cells. *Traffic, 9*(2), 230–250. https://doi.org/10.1111/j.1600-0854.2007.00677.x.

Feng, Q., Stork, C. J., Xu, S., Yuan, D., Xia, X., LaPenna, K. B., et al. (2019). Increased circulating microparticles in streptozotocin-induced diabetes propagate inflammation contributing to microvascular dysfunction. *The Journal of Physiology, 597*(3), 781–798. https://doi.org/10.1113/JP277312.

Futter, C. E., Pearse, A., Hewlett, L. J., & Hopkins, C. R. (1996). Multivesicular endosomes containing internalized EGF-EGF receptor complexes mature and then fuse directly with lysosomes. *The Journal of Cell Biology*, *132*(6), 1011–1023. https://doi.org/10.1083/jcb.132.6.1011.

Galione, A. (2006). NAADP, a new intracellular messenger that mobilizes Ca2+ from acidic stores. *Biochemical Society Transactions*, *34*(Pt 5), 922–926. https://doi.org/10.1042/BST0340922.

Glunde, K., Guggino, S. E., Solaiyappan, M., Pathak, A. P., Ichikawa, Y., & Bhujwalla, Z. M. (2003). Extracellular acidification alters lysosomal trafficking in human breast cancer cells. *Neoplasia*, *5*(6), 533–545. https://doi.org/10.1016/s1476-5586(03)80037-4.

Guse, A. H. (2012). Linking NAADP to ion channel activity: A unifying hypothesis. *Science Signaling*, *5*(221), pe18. https://doi.org/10.1126/scisignal.2002890.

Hasegawa, T., Konno, M., Baba, T., Sugeno, N., Kikuchi, A., Kobayashi, M., et al. (2011). The AAA-ATPase VPS4 regulates extracellular secretion and lysosomal targeting of alpha-synuclein. *PLoS One*, *6*(12), e29460. https://doi.org/10.1371/journal.pone.0029460.

Hessvik, N. P., & Llorente, A. (2018). Current knowledge on exosome biogenesis and release. *Cellular and Molecular Life Sciences*, *75*(2), 193–208. https://doi.org/10.1007/s00018-017-2595-9.

Hoglinger, D., Haberkant, P., Aguilera-Romero, A., Riezman, H., Porter, F. D., Platt, F. M., et al. (2015). Intracellular sphingosine releases calcium from lysosomes. *eLife*, *4*, e10616. https://doi.org/10.7554/eLife.10616.

Jahreiss, L., Menzies, F. M., & Rubinsztein, D. C. (2008). The itinerary of autophagosomes: From peripheral formation to kiss-and-run fusion with lysosomes. *Traffic*, *9*(4), 574–587. https://doi.org/10.1111/j.1600-0854.2008.00701.x.

Ju, R., Zhuang, Z. W., Zhang, J., Lanahan, A. A., Kyriakides, T., Sessa, W. C., et al. (2014). Angiopoietin-2 secretion by endothelial cell exosomes: Regulation by the phosphatidylinositol 3-kinase (PI3K)/Akt/endothelial nitric oxide synthase (eNOS) and syndecan-4/syntenin pathways. *The Journal of Biological Chemistry*, *289*(1), 510–519. https://doi.org/10.1074/jbc.M113.506899.

Kinnear, N. P., Wyatt, C. N., Clark, J. H., Calcraft, P. J., Fleischer, S., Jeyakumar, L. H., et al. (2008). Lysosomes co-localize with ryanodine receptor subtype 3 to form a trigger zone for calcium signalling by NAADP in rat pulmonary arterial smooth muscle. *Cell Calcium*, *44*(2), 190–201. https://doi.org/10.1016/j.ceca.2007.11.003.

Kobayashi, M., Inoue, K., Warabi, E., Minami, T., & Kodama, T. (2005). A simple method of isolating mouse aortic endothelial cells. *Journal of Atherosclerosis and Thrombosis*, *12*(3), 138–142. https://doi.org/10.5551/jat.12.138.

Lazaro-Ibanez, E., Sanz-Garcia, A., Visakorpi, T., Escobedo-Lucea, C., Siljander, P., Ayuso-Sacido, A., et al. (2014). Different gDNA content in the subpopulations of prostate cancer extracellular vesicles: Apoptotic bodies, microvesicles, and exosomes. *Prostate*, *74*(14), 1379–1390. https://doi.org/10.1002/pros.22853.

Li, X., Han, W. Q., Boini, K. M., Xia, M., Zhang, Y., & Li, P. L. (2013). TRAIL death receptor 4 signaling via lysosome fusion and membrane raft clustering in coronary arterial endothelial cells: Evidence from ASM knockout mice. *Journal of Molecular Medicine (Berlin, Germany)*, *91*(1), 25–36. https://doi.org/10.1007/s00109-012-0968-y.

Li, G., Huang, D., Bhat, O. M., Poklis, J. L., Zhang, A., Zou, Y., et al. (2020). Abnormal podocyte TRPML1 channel activity and exosome release in mice with podocyte-specific Asah1 gene deletion. *Biochimica et Biophysica Acta - Molecular and Cell Biology of Lipids*, *1866*(2), 158856. https://doi.org/10.1016/j.bbalip.2020.158856.

Li, G., Huang, D., Hong, J., Bhat, O. M., Yuan, X., & Li, P. L. (2019). Control of lysosomal TRPML1 channel activity and exosome release by acid ceramidase in mouse podocytes. *American Journal of Physiology. Cell Physiology*, *317*(3), C481–C491. https://doi.org/10.1152/ajpcell.00150.2019.

Li, G., Huang, D., Li, N., Ritter, J. K., & Li, P. L. (2021). Regulation of TRPML1 channel activity and inflammatory exosome release by endogenously produced reactive oxygen species in mouse podocytes. *Redox Biology*, *43*, 102013. https://doi.org/10.1016/j.redox.2021.102013.

Li, G., Kidd, J., Kaspar, C., Dempsey, S., Bhat, O. M., Camus, S., et al. (2020). Podocytopathy and nephrotic syndrome in mice with podocyte-specific deletion of the Asah1 gene: Role of ceramide accumulation in glomeruli. *The American Journal of Pathology*, *190*(6), 1211–1223. https://doi.org/10.1016/j.ajpath.2020.02.008.

Li, X., Rydzewski, N., Hider, A., Zhang, X., Yang, J., Wang, W., et al. (2016). A molecular mechanism to regulate lysosome motility for lysosome positioning and tubulation. *Nature Cell Biology*, *18*(4), 404–417. https://doi.org/10.1038/ncb3324.

Li, P. L., Zhang, Y., Abais, J. M., Ritter, J. K., & Zhang, F. (2013). Cyclic ADP-ribose and NAADP in vascular regulation and diseases. *Messenger (Los Angel)*, *2*(2), 63–85. https://doi.org/10.1166/msr.2013.1022.

Lin, S. X., & Collins, C. A. (1992). Immunolocalization of cytoplasmic dynein to lysosomes in cultured cells. *Journal of Cell Science*, *101*(Pt 1), 125–137. https://doi.org/10.1242/jcs.101.1.125.

Masuko, K., Murata, M., Suematsu, N., Okamoto, K., Yudoh, K., Nakamura, H., et al. (2009). A metabolic aspect of osteoarthritis: Lipid as a possible contributor to the pathogenesis of cartilage degradation. *Clinical and Experimental Rheumatology*, *27*(2), 347–353.

Mullock, B. M., Bright, N. A., Fearon, C. W., Gray, S. R., & Luzio, J. P. (1998). Fusion of lysosomes with late endosomes produces a hybrid organelle of intermediate density and is NSF dependent. *The Journal of Cell Biology*, *140*(3), 591–601. https://doi.org/10.1083/jcb.140.3.591.

Murrow, L., & Debnath, J. (2015). ATG12-ATG3 connects basal autophagy and late endosome function. *Autophagy*, *11*(6), 961–962. https://doi.org/10.1080/15548627.2015.1040976.

Nozaki, T., Sugiyama, S., Koga, H., Sugamura, K., Ohba, K., Matsuzawa, Y., et al. (2009). Significance of a multiple biomarkers strategy including endothelial dysfunction to improve risk stratification for cardiovascular events in patients at high risk for coronary heart disease. *Journal of the American College of Cardiology*, *54*(7), 601–608. https://doi.org/10.1016/j.jacc.2009.05.022.

Perrotta, I., & Aquila, S. (2016). Exosomes in human atherosclerosis: An ultrastructural analysis study. *Ultrastructural Pathology*, *40*(2), 101–106. https://doi.org/10.3109/01913123.2016.1154912.

Piccoli, E., Nadai, M., Caretta, C. M., Bergonzini, V., Del Vecchio, C., Ha, H. R., et al. (2011). Amiodarone impairs trafficking through late endosomes inducing a Niemann-pick C-like phenotype. *Biochemical Pharmacology*, *82*(9), 1234–1249. https://doi.org/10.1016/j.bcp.2011.07.090.

Pu, J., Guardia, C. M., Keren-Kaplan, T., & Bonifacino, J. S. (2016). Mechanisms and functions of lysosome positioning. *Journal of Cell Science*, *129*(23), 4329–4339. https://doi.org/10.1242/jcs.196287.

Rajagopal, C., & Harikumar, K. B. (2018). The origin and functions of exosomes in Cancer. *Frontiers in Oncology*, *8*, 66. https://doi.org/10.3389/fonc.2018.00066.

Schieder, M., Rotzer, K., Bruggemann, A., Biel, M., & Wahl-Schott, C. (2010). Planar patch clamp approach to characterize ionic currents from intact lysosomes. *Science Signaling*, *3*(151), pl3. https://doi.org/10.1126/scisignal.3151pl3.

Schuchman, E. H. (2016). Acid ceramidase and the treatment of ceramide diseases: The expanding role of enzyme replacement therapy. *Biochimica et Biophysica Acta*, *1862*(9), 1459–1471. https://doi.org/10.1016/j.bbadis.2016.05.001.

Shen, D., Wang, X., Li, X., Zhang, X., Yao, Z., Dibble, S., et al. (2012). Lipid storage disorders block lysosomal trafficking by inhibiting a TRP channel and lysosomal calcium release. *Nature Communications*, *3*, 731. https://doi.org/10.1038/ncomms1735.

Sikora, J., Dworski, S., Jones, E. E., Kamani, M. A., Micsenyi, M. C., Sawada, T., et al. (2017). Acid ceramidase deficiency in mice results in a broad range of central nervous system abnormalities. *The American Journal of Pathology, 187*(4), 864–883. https://doi.org/10.1016/j.ajpath.2016.12.005.

Sinning, J. M., Losch, J., Walenta, K., Bohm, M., Nickenig, G., & Werner, N. (2011). Circulating CD31+/Annexin V+ microparticles correlate with cardiovascular outcomes. *European Heart Journal, 32*(16), 2034–2041. https://doi.org/10.1093/eurheartj/ehq478.

Subra, C., Grand, D., Laulagnier, K., Stella, A., Lambeau, G., Paillasse, M., et al. (2010). Exosomes account for vesicle-mediated transcellular transport of activatable phospholipases and prostaglandins. *Journal of Lipid Research, 51*(8), 2105–2120. https://doi.org/10.1194/jlr.M003657.

Trajkovic, K., Dhaunchak, A. S., Goncalves, J. T., Wenzel, D., Schneider, A., Bunt, G., et al. (2006). Neuron to glia signaling triggers myelin membrane exocytosis from endosomal storage sites. *The Journal of Cell Biology, 172*(6), 937–948. https://doi.org/10.1083/jcb.200509022.

Vingtdeux, V., Hamdane, M., Loyens, A., Gele, P., Drobeck, H., Begard, S., et al. (2007). Alkalizing drugs induce accumulation of amyloid precursor protein by-products in luminal vesicles of multivesicular bodies. *The Journal of Biological Chemistry, 282*(25), 18197–18205. https://doi.org/10.1074/jbc.M609475200.

Ward, D. M., Pevsner, J., Scullion, M. A., Vaughn, M., & Kaplan, J. (2000). Syntaxin 7 and VAMP-7 are soluble N-ethylmaleimide-sensitive factor attachment protein receptors required for late endosome-lysosome and homotypic lysosome fusion in alveolar macrophages. *Molecular Biology of the Cell, 11*(7), 2327–2333. https://doi.org/10.1091/mbc.11.7.2327.

Wei, Y. M., Li, X., Xu, M., Abais, J. M., Chen, Y., Riebling, C. R., et al. (2013). Enhancement of autophagy by simvastatin through inhibition of Rac1-mTOR signaling pathway in coronary arterial myocytes. *Cellular Physiology and Biochemistry, 31*(6), 925–937. https://doi.org/10.1159/000350111.

Xu, M., Li, X. X., Xiong, J., Xia, M., Gulbins, E., Zhang, Y., et al. (2013). Regulation of autophagic flux by dynein-mediated autophagosomes trafficking in mouse coronary arterial myocytes. *Biochimica et Biophysica Acta, 1833*(12), 3228–3236. https://doi.org/10.1016/j.bbamcr.2013.09.015.

Yuan, X., Bhat, O. M., Lohner, H., Zhang, Y., & Li, P. L. (2019). Endothelial acid ceramidase in exosome-mediated release of NLRP3 inflammasome products during hyperglycemia: Evidence from endothelium-specific deletion of Asah1 gene. *Biochimica et Biophysica Acta - Molecular and Cell Biology of Lipids, 1864*(12), 158532. https://doi.org/10.1016/j.bbalip.2019.158532.

Zhang, F., Jin, S., Yi, F., & Li, P. L. (2009). TRP-ML1 functions as a lysosomal NAADP-sensitive Ca2+ release channel in coronary arterial myocytes. *Journal of Cellular and Molecular Medicine, 13*(9B), 3174–3185. https://doi.org/10.1111/j.1582-4934.2008.00486.x.

Zhang, F., & Li, P. L. (2007). Reconstitution and characterization of a nicotinic acid adenine dinucleotide phosphate (NAADP)-sensitive Ca2+ release channel from liver lysosomes of rats. *The Journal of Biological Chemistry, 282*(35), 25259–25269. https://doi.org/10.1074/jbc.M701614200.

Zhang, F., Xia, M., & Li, P. L. (2010). Lysosome-dependent Ca(2+) release response to Fas activation in coronary arterial myocytes through NAADP: Evidence from CD38 gene knockouts. *American Journal of Physiology. Cell Physiology, 298*(5), C1209–C1216. https://doi.org/10.1152/ajpcell.00533.2009.

Zhang, F., Zhang, G., Zhang, A. Y., Koeberl, M. J., Wallander, E., & Li, P. L. (2006). Production of NAADP and its role in Ca2+ mobilization associated with lysosomes in coronary arterial myocytes. *American Journal of Physiology. Heart and Circulatory Physiology, 291*(1), H274–H282. https://doi.org/10.1152/ajpheart.01064.2005.

CHAPTER FOUR

Vascular $Ca_V1.2$ channels in diabetes

Eric A. Pereira da Silva, Miguel Martín-Aragón Baudel, Junyoung Hong, Peter Bartels, Manuel F. Navedo, and Madeline Nieves-Cintrón*
Department of Pharmacology, University of California, Davis, CA, United States
*Corresponding author: e-mail address: mcnieves@ucdavis.edu

Contents

Abstract

Diabetic vasculopathy is a significant cause of morbidity and mortality in the diabetic population. Hyperglycemia, one of the central metabolic abnormalities in diabetes, has been associated with vascular dysfunction due to endothelial cell damage. However, studies also point toward vascular smooth muscle as a locus for hyperglycemia-induced vascular dysfunction. Emerging evidence implicates hyperglycemia-induced regulation of vascular L-type Ca^{2+} channels $Ca_V1.2$ as a potential mechanism for vascular dysfunction during diabetes. This chapter summarizes our current understanding of vascular $Ca_V1.2$ channels and their regulation during physiological and hyperglycemia/diabetes conditions. We will emphasize the role of $Ca_V1.2$ in vascular smooth muscle, the effects of elevated glucose on $Ca_V1.2$ function, and the mechanisms underlying its dysregulation in hyperglycemia and diabetes. We conclude by examining future directions and gaps in knowledge regarding $Ca_V1.2$ regulation in health and during diabetes.

Current Topics in Membranes, Volume 90
ISSN 1063-5823
https://doi.org/10.1016/bs.ctm.2022.09.003

1. Introduction

The World Health Organization estimates that there are about 422 million people living with diabetes worldwide and more than 95% of diabetic individuals have type 2 diabetes (WHO, 2021). Glucose homeostasis is vital for adequate bodily function. Both type 1 and type 2 diabetes impair the ability of the body to control glucose levels. Type 1 diabetes hampers insulin production by the pancreatic beta cells, whereas Type 2 diabetes limits the ability of the body to utilize insulin. Ultimately, impaired glucose homeostasis in diabetes has serious health consequences. High blood glucose level (e.g., hyperglycemia) significantly impacts the vasculature, and vascular damage is associated with increased morbidity and mortality in the diabetic population (Rangel, Rodrigues, & de Sa, 2019; Stehouwer, 2018). Diabetic hyperglycemia is associated with cognitive deficits, nephropathy, retinopathy, and neuropathy. Indeed, there is a strong association between diabetic hyperglycemia-induced damage to small resistance artery's function and these pathological conditions (Stehouwer, 2018). While endothelial dysfunction is a significant component of the diabetic vasculopathy (Fluitt, Mohit, Gambhir, & Nunlee-Bland, 2022; Shi & Vanhoutte, 2017), recent research shows that hyperglycemia-induced changes in $Ca_V1.2$ function in small diameter resistance contribute to vascular complications in diabetes (Nieves-Cintron, Flores-Tamez, Le, Baudel, & Navedo, 2021).

Small-diameter arteries and arterioles play a crucial role in regulating tissue perfusion, blood flow, and peripheral resistance. These functions are accomplished by the ability of these vessels to adjust their caliber in response to chemical and mechanical stimuli within the body. Arteries are formed by connective tissue, an inner layer of endothelial cells lining the arterial lumen, and smooth muscle cells wrapping around the vessel. Vascular smooth muscle cells harbor the contractile machinery and respond to changes in intravascular pressure by contracting. This contraction, referred to as the myogenic tone, is an intrinsic property of the vascular smooth muscle (Davis & Hill, 1999) and is regulated by the membrane potential and the magnitude of the Ca^{2+} influx through voltage-gated L-type Ca^{2+} channel $Ca_V1.2$ (Ghosh et al., 2017; Knot & Nelson, 1998; Moosmang et al., 2003).

Vascular $Ca_V1.2$ channels are the main entryway of Ca^{2+} ions into vascular smooth muscle and are essential for the myogenic tone (Moosmang et al., 2003). The activation of $Ca_V1.2$ channels in vascular smooth muscle increases global intracellular Ca^{2+} ($[Ca^{2+}]_i$), thus activating contractile proteins (e.g., excitation-contraction (EC) coupling), leading to arterial constriction (Somlyo & Somlyo, 2003). Vascular $Ca_V1.2$ channel activity also influences gene expression by regulating Ca^{2+}-driven transcriptional pathways, a process known as the excitation-transcription (ET) coupling (Nieves-Cintron, Amberg, Navedo, Molkentin, & Santana, 2008; Wamhoff, Bowles, & Owens, 2006). Thus, functional modulation of vascular $Ca_V1.2$ channels could have profound implications for arterial function by regulating contraction and gene transcription. Indeed, increased vascular $Ca_V1.2$ channel function is a well-recognized modulator of $[Ca^{2+}]_i$ and ET coupling involved in arterial remodeling and elevated myogenic tone in several pathological conditions, including hypertension and small vessel diseases (Bannister et al., 2012; Kharade et al., 2013; Navedo et al., 2008, 2010; Navedo & Amberg, 2013; Nieves-Cintron et al., 2008; Pesic, Madden, Pesic, & Rusch, 2004; Pratt, Bonnet, Ludwig, Bonnet, & Rusch, 2002; Sonkusare et al., 2006; Wang, Saada, Dai, Pang, & Palade, 2006). Emerging evidence suggests that alterations of vascular $Ca_V1.2$ channels could contribute to vascular dysfunction in diabetes (Nieves-Cintron et al., 2021). Here, we will discuss the regulation of vascular smooth muscle $Ca_V1.2$ channel during diabetes in relation to vascular function. We first provide a general view of L-type Ca^{2+} channel $Ca_V1.2$ structure and regulation and then summarize recent findings on how the channel is regulated in vascular smooth muscle and the impact on vascular function during diabetes.

2. General L-type Ca^{2+} channel $Ca_V1.2$ structure

L-type Ca^{2+} channels $Ca_V1.2$ are comprised of the pore-forming alpha subunit (α_{1C}) and auxiliary subunits β, $\alpha2\delta$, and γ that modulate channel function and trafficking (Bannister et al., 2009; Catterall, 2000). Subunit composition, receptor-mediated signaling, and binding partners provide further functional channel regulation. The $Ca_V1.2$ α_{1C} subunit contains four homologous domains (I, II, III, IV). Each domain comprises six membrane-spanning segments (S1-S6) linked by three intracellular loops and intracellular NH2- and COOH-termini (Fig. 1). Transmembrane segment 4 confers voltage sensitivity, and the ion pore is formed by the homologous S6 segments and the loop between S5 and S6 of each domain. Two glutamate residues at the P-loop determine the Ca^{2+} selectivity filter. The $Ca_V1.2$ α_{1C} subunit is the site of action of many pharmacological regulators of the channel that bind to sites within the S6 segment (Catterall, 2011) and is also post-translationally modulated by protein kinases and phosphatases

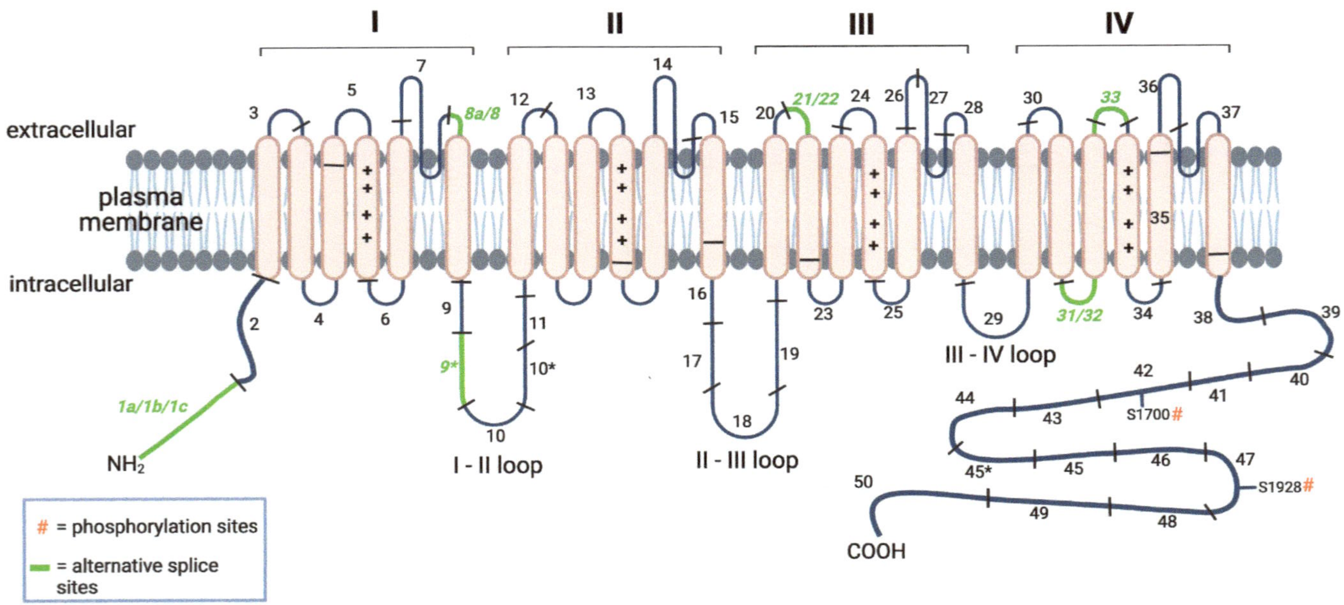

Fig. 1 Topological representation of the pore-forming α_1 subunit of the L-type calcium channel $Ca_v1.2$. The α_1 subunit comprises four homologous domains (I–IV), each consisting of six transmembrane segments (S1–S6). The positively charged S4 segment represents the voltage sensor. Numbers indicate the 50 exons encoding the α1 subunit, and black lines across the diagram represent the limits of each exon. Alternative splicing sites are shown in green. Phosphorylation sites discussed in the text are indicated by red #. Graphics created with https://BioRender.com.

(Catterall, 2000). The $Ca_V1.2$ α_{1C} subunit is also the site of action of calcium channel blockers, a mainstay therapy for hypertension (Caldwell, 1993).

The auxiliary subunits α2 and δ are the product of the same gene that gets proteolytically cleaved after translation giving rise to mature subunits connected by disulfide bonds (Catterall, 2011). The δ portion of the subunit is anchored to the plasma membrane, while the glycosylated extracellular α2 subunit interacts with the α_{1C} subunit (Gurnett, De Waard, & Campbell, 1996). Knockdown of the α2δ1 isoform in rat cerebral arteries reduces Ca^{2+} influx through vascular $Ca_V1.2$ channels leading to artery dilation (Bannister et al., 2009). Conversely, higher α2δ1 expression has been associated with increased vascular $Ca_V1.2$ surface membrane expression, currents, tone, and hypertension (Bannister et al., 2012). These observations suggest a role for the α2δ subunit in controlling vascular $Ca_V1.2$ trafficking, surface abundance, and activity, thus influencing vascular function in health and disease.

The auxiliary β subunit is a cytoplasmic protein consisting of two conserved core regions akin to the Src-homology-3 (SH3) domain and the guanylate kinase domain (Dolphin, 2003). The two core regions are connected by a HOOK domain and flanked by NH_2 and COOH-terminus. The β subunit interacts with the alpha subunit by binding to a conserved Alpha Interaction Domain (AID) at the intracellular linker loop between the homologous domains I and II of the α_{1C} subunit (Dolphin, 2016; Van Petegem, Clark, Chatelain, & Minor, 2004). The β subunit enhances the expression, stability, and trafficking of the α_{1C} subunit and modulates the channel's biophysical properties (Dolphin, 2016; Takahashi, Miriyala, Tay, Yue, & Colecraft, 2005). Vascular smooth muscle cells express β2 and β3 subunit (Cox & Fromme, 2015; Kharade et al., 2013; Reimer, Huber, Garcia, Haase, & Striessnig, 2000). Recent findings have shown that the β3 subunit expression is prominent in mesenteric vascular smooth muscle and is upregulated and contributes to enhanced vascular $Ca_V1.2$ current density during angiotensin II (AngII)-induced hypertension (Kharade et al., 2013). The β3-dependent changes in vascular $Ca_V1.2$ activity during hypertension were correlated with changes in arterial tone and blood pressure, consistent with altered vascular function during this pathological condition. These results highlight a prominent role for auxiliary β subunits, particularly β3, in controlling $Ca_V1.2$ activity and vascular function during physiological and pathological conditions. It is tempting to speculate that both α2δ1 and β3 subunits could be potential new druggable targets to treat cardiovascular diseases such as hypertension.

There are eight known γ subunit isoforms (γ1–γ8), each formed by four transmembrane proteins with intracellular amino and carboxy termini in addition to a common GLWXXC amino acid motif in the first loop (Arikkath & Campbell, 2003; Burgess, Gefrides, Foreman, & Noebels, 2001). Although the functional role of the γ subunits may vary within different tissues, at the moment, it is mainly associated with the regulation of the biophysical properties of the cardiac $Ca_V1.2$ channel (Arikkath & Campbell, 2003). For instance, in rat and human hearts, it was demonstrated that expression of the isoforms γ4, γ6, γ7, and γ8 modulated the activity of $Ca_V1.2$ channels (Yang, Katchman, Morrow, Doshi, & Marx, 2011). However, the functional relevance of the γ subunit in vascular $Ca_V1.2$ channels and the implications for vascular function are still unknown, which may represent an opportunity for future work.

Protein kinases and phosphatases also regulate L-type Ca^{2+} channels $Ca_V1.2$. In addition, alternative splicing of mRNA encoding L-type Ca^{2+} channels $Ca_V1.2$ subunits gives rise to tissue-selective $Ca_V1.2$ channels with distinct pharmacological and electrophysiological profiles, thus representing an additional layer of regulation that increases channel functional diversity (Catterall, 2000; Cheng et al., 2009). In Sections 3.1 and 3.2 we discuss L-type Ca^{2+} channels $Ca_V1.2$ channel regulation by post-transcriptional and translational modifications that influence channel function in physiology and pathology, emphasizing vascular $Ca_V1.2$.

3. Regulation of vascular $Ca_V1.2$ channels

3.1 Splice variants

$Ca_V1.2$ channels are subject to alternative splicing—a molecular biological transcriptional process that generates many variants from a single gene and may be involved in functional $Ca_V1.2$ diversity in health and disease (Hu, Liang, & Soong, 2017; Liao & Soong, 2010; Liao et al., 2007; Yang et al., 2000). The human gene for the $Ca_V1.2$ channel (cacna1c) has 50 exons, whereas the mouse and rat forms have 49 exons (Cheng et al., 2007; Liao, Yong, Liang, Yue, & Soong, 2005; Soldatov, 1994). At least 20 exons in the human cacna1c gene are possible sites for alternative splicing. $Ca_V1.2$ alternative splices have been extensively studied, especially in the heart and smooth muscle, due to their versatility in function, localization, and implications in cardiovascular diseases (Bannister et al., 2011; Hu et al., 2017; Liao & Soong, 2010; Yang et al., 2000). Cardiac myocytes and vascular smooth muscle are termed $Ca_V1.2a$ and $Ca_V1.2b$, respectively (Liao & Soong, 2010).

The smooth muscle isoform of $Ca_V1.2$ is different from the cardiac counterpart in several regions, including mutually exclusive exons 1a/1b in the NH_2 terminal, at the first homologous domain (IS6), which is coded by mutually exclusive exon 8a in the heart, and exon 8 in smooth muscle, and at the S3 of the fourth homologous domain (IVS3), which contains exon 31 or 32 in the cardiac and smooth muscle $Ca_V1.2$, respectively (Liao et al., 2005; Tang, Hong, Wang, & Soong, 2007; Yeow, Loh, & Soong, 2021). The smooth muscle $Ca_V1.2b$ isoform also contains exon 9*, an insertion of 25 amino acid residues at the cytosolic loop between homologous domains I and II (Cheng et al., 2009; Liao et al., 2005, 2007) (Fig. 1). Thus, cardiac $Ca_V1.2a$ has exons 1a/8a/-9*/31/33, while smooth muscle $Ca_V1.2b$ comprises exons 1b/8/9*/32/33 (Fig. 1). Splice variations give rise to $Ca_V1.2$ channels with distinct functional and pharmacological characteristics. For instance, the smooth muscle $Ca_V1.2b$ channels are more sensitive to blockade by dihydropyridines (DHPs) than the cardiac channels (Hu et al., 2017; Hu & Marban, 1998; Liao et al., 2007; Triggle, 1991). The transmembrane segment IS6 determines the binding affinity of DHPs. Thus, tissue-specific expression of mutually exclusive exon 8 in IS6 of the $Ca_V1.2b$ isoform is partly responsible for the enhanced sensitivity to DHPs in smooth muscle channels (Liao & Soong, 2010; Welling et al., 1997). A splice variation in human $Ca_V1.2b$ that excludes exon 33 ($Ca_V1.2_{SM}$) has window currents closer to membrane resting potential and is more sensitive to the nifedipine blockade (Cheng et al., 2007; Liao et al., 2004, 2007; Tang et al., 2004). Calcium channel blockers such as DHPs are the mainstay therapies in treating cardiovascular complications such as arrhythmias, angina, and hypertension. Thus, tissue-specific expression of $Ca_V1.2$ variants could be further leveraged to develop tissue-specific calcium channel blockers with increased sensitivity to avoid unwanted side effects.

$Ca_V1.2$ channels containing the splice variant 9*, which exhibit a hyperpolarizing shift in the activation potential, have been suggested to be dominant and functionally critical in cerebral smooth muscle cells for control of Ca^{2+} influx, artery diameter, and blood flow (Nystoriak, Murakami, Penar, & Wellman, 2009). However, full-length α_{1C} subunits were found expressed in small resistance cerebral arteries (Cheng et al., 2007). Authors found that this smooth muscle-specific α_{1C} subunit has a splice variation in exon 1 (called exon 1c), which gives rise to splice variant $Ca_V1.2e1c$. This variant has a cysteine-rich N-terminus. When expressed with $\alpha 2\delta$ subunits, the $Ca_V1.2e1c$ channel displays steady-state half activation and deactivation at more negative potentials, smaller whole-cell current density, and decreased plasma membrane surface abundance (Cheng et al., 2007). While additional experiments should be carried out further to define the role of smooth muscle-specific $Ca_V1.2$ variants, the diversity of splice variants within the same tissue could provide specific regulation of vascular physiology and likely be distinctly regulated in health and disease states.

Alterations in the expression profile of the $Ca_V1.2$ splice variants have been detected in cardiovascular conditions such as heart failure, atherosclerosis, and hypertension (Hu et al., 2017). For example, mutually exclusive splicing of membrane-spanning segment S3 of homologous domain IV (IVS3) in cardiac $Ca_V1.2$ is developmentally regulated (Diebold et al., 1992). While both isoforms IVS3A (exon 31) and IVS3B (exon 32) are found in rat fetal cardiomyocytes, the IVS3B isoform predominates in adult rat cardiomyocytes (Gidh-Jain, Huang, Jain, Battula, & el-Sherif, N., 1995). *Re*-emergence of the fetal isoform was reported after myocardial infarction in rat hearts (Gidh-Jain et al., 1995). Interestingly in non-failing human hearts, IVS3A was found to be the predominant isoform, and a shift to higher expression of IVS3B was observed in failing human hearts (Yang et al., 2000). In addition, several studies have reported increased expression of $Ca_V1.2$ channels containing the 9* splice variant in cardiac tissue from rodent hypertrophy models and myocardial infarction (Fan et al., 2019; Liao et al., 2009; Tang et al., 2008). Intriguingly, changes in α_{1C} splice variants were prominent in areas close to the infarct region compared to areas away from the damaged zone (Liao et al., 2009). It will be intriguing to determine if such variations in α_{1C} splice variants also occur in regions of blood vessels that are close to lesions, such as atherosclerotic plaques, compared to other areas. Regardless, alteration in $Ca_V1.2$ splice variants is also noted in atherosclerosis where exon 9* is decreased, and a switch to increased expression of mutually exclusive exon 22 instead of 21 was detected in arterial tissue from atherosclerotic plaque (Tiwari, Zhang, Heller, Abernethy, & Soldatov, 2006). A diverse expression profile of splice exons has also been described in the heart of rat models of genetic hypertension (Tang et al., 2008). It is unclear whether diabetic hyperglycemia impacts the expression profile of $Ca_V1.2$ splice variants in vascular smooth muscle. Pathological changes in the expression pattern of splice variants could lead to changes in channel function (Gidh-Jain et al., 1995; Hu et al., 2017; Liao et al., 2007; Tiwari et al., 2006) with a potential to impact organ physiology and influence regulation by cellular signaling. Accordingly, understanding diabetic hyperglycemia effects on $Ca_V1.2$ splice variants expression profile represents an area of opportunity in the vascular field.

3.2 Phosphorylation

L-type Ca^{2+} channels $Ca_V1.2$ are known to be regulated reversely by protein kinases such as protein kinases C (PKC), G (PKG), and A (PKA) (Catterall, 2000; Keef, Hume, & Zhong, 2001; Navedo & Santana, 2013; Patriarchi et al., 2016). PKC activity is typically linked with potentiation of vascular $Ca_V1.2$ channels (Amberg & Navedo, 2013; Ghosh et al., 2017; Keef et al., 2001). Indeed, research shows that vasoconstrictors like norepinephrine, endothelin, angiotensin II (angII), and serotonin mediate their actions by stimulating the vascular $Ca_V1.2$ channel activity (Keef et al., 2001; Xiong & Sperelakis, 1995). These vasoactive substances activate $G_{q/11}$ coupled receptors, which stimulate PLC activity to produce IP_3 and DAG, leading to PKC activation. The activated PKC can then stimulate vascular $Ca_V1.2$ channels leading to an increased current density (Cobine, Callaghan, & Keef, 2007; Navedo, Amberg, Nieves, Molkentin, & Santana, 2006; Seki, Yokoshiki, Sunagawa, Nakamura, & Sperelakis, 1999). The PKC-induced increase in vascular $Ca_V1.2$ current density is presumably mediated by phosphorylation of the α_{1C} subunit or an accessory subunit, although this hypothesis requires confirmation. PKC effects on vascular $Ca_V1.2$ channels and vascular function require anchoring of the kinase near the α_{1C} subunit by the anchoring protein AKAP5. Accordingly, genetic ablation of AKAP5 or disruption of the interaction between AKAP5 and PKC prevents AngII-induced potentiation of vascular $Ca_V1.2$ current density and vasoconstriction (Navedo et al., 2008; Nystoriak et al., 2017). The role of AKAP5 in this complex also seems to be essential for controlling blood pressure as AKAP5 knockout mice are hypotensive (Navedo et al., 2008).

cGMP-dependent protein kinase-1 (PKG1) is abundant in vascular smooth muscle and is recognized as an important vasodilation mediator (Francis, Noblett, Todd, Wells, & Corbin, 1988; Weber et al., 2007). One of the notable physiological events carried through this pathway is the action of NO and natriuretic peptides. Both paracrine stimulators activate guanylyl cyclase in VSM to produce cGMP and activate PKG. PKG, in turn, acts on several cellular targets and induces vasodilation in part by decreasing the intracellular calcium (Carvajal, Germain, Huidobro-Toro, & Weiner, 2000; Keef et al., 2001). Some studies have reported that PKG decreases vascular $Ca_V1.2$ activity, which may be one of the mechanisms contributing to PKG-dependent reduction in $[Ca^{2+}]_i$ (Blatter & Wier, 1994).

PKA can regulate vascular $Ca_V1.2$ channels, but the functional implications of PKA-mediated regulation of vascular $Ca_V1.2$ have been difficult to establish since some studies suggest that PKA has a blunting effect on the channel activity (Xiong & Sperelakis, 1995; Xiong, Sperelakis, & Fenoglio-Preiser, 1994), while others indicate potentiation of channel activity (Keef et al., 2001; Ruiz-Velasco, Zhong, Hume, & Keef, 1998; Tewari & Simard, 1994). For example, whole-cell experiments in which the PKA-catalytic subunit was dialyzed into the cells through the patch pipette decreased $Ca_V1.2$ current density in rabbit portal vein smooth muscle (Xiong et al., 1994). Moreover, application of the cell-wide activator of endogenous adenylate cyclase (AC) forskolin and membrane-permeable cAMP analogs such as 8-Br-derivatives were reported to inhibit $Ca_V1.2$ channels in rabbit portal vein smooth muscle (Xiong et al., 1994; Xiong & Sperelakis, 1995). Conversely, similar stimuli (e.g., beta-adrenergic agonists, forskolin, the cAMP analog 8-Br-cAMP, and PKA catalytic subunit) have been reported to stimulate $Ca_V1.2$ current density in guinea-pig basilar artery and rabbit portal vein smooth muscle (Keef et al., 2001; Ruiz-Velasco et al., 1998; Tewari & Simard, 1994). Intriguingly, low intracellular cAMP levels in rabbit portal vein smooth muscle induced an increase in Ba^{2+} current through the $Ca_V1.2$ channel, but increasing cAMP levels blunted $Ca_V1.2$-mediated currents (Ishikawa, Hume, & Keef, 1993). Interestingly, the study by Syed et al. showed that elevating extracellular glucose causes a slight albeit significant increase in cAMP, leading to PKA-dependent potentiation of $Ca_V1.2$ currents and vasoconstriction (Syed et al., 2019). On the other hand, forskolin elevated cell-wide cAMP levels leading to vasodilation, even in the presence of glucose (Syed et al., 2019). Thus, low cAMP concentrations may lead to PKA-dependent potentiation of $Ca_V1.2$ activity and contraction, while high cAMP levels may lead to activation of vasodilatory pathways that override the $Ca_V1.2$-mediated Ca^{2+} influx.

Because of the distinct modulation of vascular $Ca_V1.2$ channels by PKA, the physiological implications of this regulation have been challenging to establish. However, recent work found that PKA is necessary for potentiation of mouse and rat cerebral, mouse mesenteric, and mouse femoral vascular smooth muscle in response to acute elevations in extracellular glucose (e.g., hyperglycemia) and in diabetes (Navedo, Takeda, Nieves-Cintron, Molkentin, & Santana, 2010; Nystoriak et al., 2017; Syed et al., 2019). The glucose-induced, PKA-dependent potentiation of vascular $Ca_V1.2$ channels was correlated with an increase in cerebral artery constriction

(Navedo, Takeda, et al., 2010; Nystoriak et al., 2017). Because of the novelty of this finding, the following section summarizes recent studies on the effects and mechanisms underlying alterations in vascular $Ca_V1.2$ channel function during hyperglycemia and diabetes and the influence of these changes on vascular reactivity.

4. Effects of hyperglycemia on $Ca_V1.2$

$Ca_v1.2$ is differentially regulated during hyperglycemia and diabetes in different vascular beds, with studies reporting no change, decrease, or potentiation of $Ca_V1.2$ activity. For instance, a study using a streptozotocin (STZ) type 1 diabetic model showed decreased $Ca_V1.2$ current density in VSM from rat tail artery as the diabetic phenotype progressed from 1 week to 3 months. A swine model of diabetes (hyperglycemia only) and diabetic dyslipidemia decreased L-type Ca^{2+} current density in VSM of the coronary conduit artery (Witczak, Wamhoff, & Sturek, 2006). Kidney arterioles from (STZ)-induced diabetic rats showed reduced contractile responses to $Ca_V1.2$ agonist Bay K 8644 and K^+-induced depolarization, suggesting a reduced $Ca_V1.2$ function (Carmines, Ohishi, & Ikenaga, 1996). The myogenic tone of cerebral arteries from diabetic Goto-Kakizaki (GK) rats was reduced compared to control Wistar-Kyoto (WKY) rats. This was attributed to a decreased sensitivity to calcium due to a defective Rho-kinase signaling pathway (Abd-Elrahman et al., 2017; Kold-Petersen, Brondum, Nilsson, Flyvbjerg, & Aalkjaer, 2012).

Many studies, however, report significant increases in channel activity and vascular reactivity in arterial myocytes and arteries during hyperglycemia and diabetes. Aortic VSM from STZ-treated mice showed a higher density of $Ca_V1.2$ currents and augmented contractility to phenylephrine compared to control mice. The increase in $Ca_V1.2$ activity was linked to increased expression of PI3Kδ (Pinho et al., 2010). Mesenteric arteries from STZ-treated rats showed enhanced contractility to α-adrenergic agonists that were dependent on extracellular calcium concentration; moreover, high potassium and the $Ca_V1.2$ agonist Bay K 8644 elicited higher contractile responses in these arteries than in control arteries (White & Carrier, 1990), suggesting enhanced activation of $Ca_V1.2$. Similarly, myogenic tone and contraction to the $Ca_V1.2$ agonist, Bay K 8644, were enhanced in skeletal muscle arterioles from STZ-diabetic rats. The study also showed increased responsiveness to phenylephrine and the PKC activator PMA; thus, the authors suggested that enhanced activation of $Ca_V1.2$ and PKC in arteriolar smooth muscle is responsible for increased myogenic tone (Ungvari et al., 1999). A recent study using male diabetic GK rats showed augmented $Ca_V1.2$ current density in VSM from cerebral arteries compared to WKY controls (Ma et al., 2017). The study reported higher mRNA and protein expression levels of the $Ca_V1.2\alpha$ subunit in GK's arteries compared to the WKY controls (Ma et al., 2017). Acute exposure to a hyperglycemic stimulus potentiated $Ca_V1.2$ activity in cerebral artery VSM, requiring PKA activity (Navedo, Takeda, et al., 2010). Further studies showed increased phosphorylation of the putative phosphorylation site serine 1928 (S1928) (Nystoriak et al., 2017). Noteworthy, increased phosphorylation of S1928 was also observed in two murine models of diabetes (high-fat diet and STZ) (Syed et al., 2019) and arterial tissue from diabetic patients (Nystoriak et al., 2017; Prada et al., 2019, 2020; Syed et al., 2019), whole-cell $Ca_V1.2$ currents density was robustly increased in VSM from these diabetic models and humans with diabetes. Moreover, in human arterial myocytes from nondiabetic patients, incubation of 15 mM D-glucose increased $Ca_V1.2$ current density and elevated phosphorylation of S1928 (Nystoriak et al., 2017). At the organ level, increased $Ca_V1.2$ activity in resulted in significantly enhanced vasoconstriction and myogenic tone in arteries exposed to high glucose and arteries isolated from diabetes models (Ma et al., 2017; Navedo, Cheng, et al., 2010; Nystoriak et al., 2017; Prada et al., 2019, 2020). The reported differences could represent diversity in regulating $Ca_V1.2$ channels in different vascular beds and animal models. The severity of the hyperglycemic state could also impact these results.

5. Mechanisms of hyperglycemia-induced regulation of vascular $Ca_V1.2$ channels

The mechanisms underlying the potentiation of vascular $Ca_V1.2$ channels in diabetes have been the subject of intense investigation. In the last few years, a series of studies identified and characterized a molecular nanocomplex involving AKAP5, $P2Y_{11}$, AC5, and PKA as the final effector of vascular $Ca_V1.2$ activation in response to diabetic hyperglycemia (Prada et al., 2019, 2020; Syed et al., 2019). Subcellular compartmentalization of signaling domains provides a means to fine-tune signal transduction by keeping signaling molecules in proximity (Langeberg & Scott, 2005). Compartmentalization of second messengers such as Ca^{2+} and cAMP, as well as signaling proteins, is essential to control the subcellular spatiotemporal

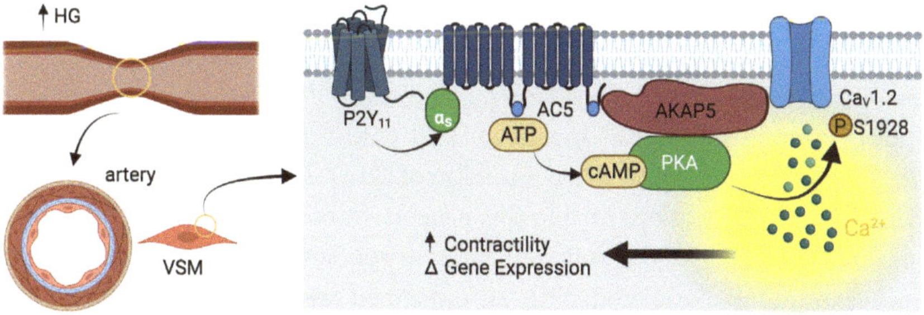

Fig. 2 Proposed model for $Ca_V1.2$ channel regulation in hyperglycemia/diabetes. Diabetes engages a signaling complex facilitated by AKAP5 that includes the purinergic receptor $P2Y_{11}$, adenylyl cyclase V, PKA, and $Ca_V1.2$ channels. During hyperglycemic conditions, glucose metabolism leads to the production of nucleotides and activation of the Gs-coupled purinergic receptor $P2Y_{11}$, which in turn activates AC5, increasing cAMP and PKA locally. AKAP5 keeps PKA and $Ca_V1.2$ in proximity, thus facilitating phosphorylation of the $Ca_V1.2$ at S1928, which potentiates channel activity. Increased $Ca_V1.2$ activity increases intracellular Ca^{2+} leading to VSM contraction. Graphics created with BioRender.com.

response to different environmental stimuli (Gallegos, Kunkel, & Newton, 2006; Halls & Cooper, 2011; Houslay, 2010; Vaca, 2010). In the vascular smooth muscle, scaffold proteins such as A-kinase anchoring proteins (AKAPs) provide a platform to help organize the $P2Y_{11}$/AC5/cAMP/PKA/$Ca_V1.2$ signaling domain (Fig. 2). In this section, we discuss the evidence for the engagement of this molecular nanocomplex in the activation of vascular $Ca_V1.2$ channels during diabetic hyperglycemia and discuss the involvement of the different components of this axis in increasing vascular reactivity during diabetes.

5.1 PKA

The description of the role of each member of the $P2Y_{11}$/AC5/cAMP/PKA signaling axis (Fig. 2) on the regulation of vascular $Ca_V1.2$ channels in response to diabetic hyperglycemia starts with PKA as this was the first component in this new complex that was identified. Although PKC activity is increased in vascular smooth muscle cells during diabetic hyperglycemia (Hashim, Li, & Anand-Srivastava, 2006; Patel et al., 1999; Williams, Gallacher, Patel, & Orme, 1997; Williams & Schrier, 1992) and this kinase is required for enhanced vascular $Ca_V1.2$ activity during hypertension (Navedo et al., 2006; Navedo, Amberg, Votaw, & Santana, 2005; Nieves-Cintron et al., 2008; Payne & Smeda, 2002; Ruan & Arendshorst, 1996; Salamanca & Khalil, 2005), elevations in extracellular glucose were still able to stimulate channel activity in cerebral vascular smooth muscle from PKC knockout mice (Navedo, Takeda, et al., 2010; Nystoriak et al., 2017). However, inhibiting PKA in cerebral vascular smooth muscle from PKC knockout and wild-type mice was sufficient to prevent the glucose-induced potentiation of $Ca_V1.2$ channels, thus suggesting that PKA activity was essential in this process. Subsequent experiments demonstrated that the glucose-induced PKA-dependent potentiation of vascular $Ca_V1.2$ channels correlated with an enhanced vasoconstriction (Nystoriak et al., 2017; Prada et al., 2019, 2020; Syed et al., 2019). This discovery was paradoxical as PKA activity has traditionally been linked to vasorelaxation (Aiello, Walsh, & Cole, 1995; Moore, Nelson, Parelkar, Rusch, & Rhee, 2014; Porter et al., 1998; Wellman, Bonev, Nelson, & Brayden, 1996; Wellman, Santana, Bonev, & Nelson, 2001). Thus, a key question from this work was how glucose-induced PKA activity leads to vasoconstriction. The current model suggests that elevations in extracellular glucose likely activate a subset of PKA molecules close to vascular $Ca_V1.2$ channels (Fig. 2). This arrangement would facilitate stimulation of vascular $Ca_V1.2$ channels leading to elevations in global $[Ca^{2+}]_i$ and vasoconstriction while sparing the activation of other PKA-dependent pathways mainly associated with vasorelaxation.

Another important question from the work described above is how glucose-induced PKA activity stimulates vascular $Ca_V1.2$ function? Several PKA phosphorylation sites have been described in $Ca_V1.2$ channels (Catterall, 2000). Earlier studies identified serine 1928 (S1928) as a prominent PKA phosphorylation site in the carboxy-terminal of the α_{1C} subunit (De Jongh et al., 1996). Following this lead, Western blot experiments suggested that elevating extracellular glucose caused a significant increase in S1928 phosphorylation in human and mouse arterial lysates (Nystoriak et al., 2017). These results correlated the glucose-induced increase in S1928 phosphorylation with potentiation of $Ca_V1.2$ channels in human and mouse vascular smooth muscle and suggested that S1928 could be a key site for vascular $Ca_V1.2$ regulation during diabetic hyperglycemia. Consistent with this possibility, elevating extracellular glucose did not stimulate $Ca_V1.2$ current density in vascular smooth muscle from a mouse expressing an α_{1C} subunit in which S1928 was mutated to alanine to prevent the phosphorylation of this site (S1928A mouse) (Nystoriak et al., 2017).

Moreover, results revealed that glucose-induced vasoconstriction was prevented in S1928A mouse cerebral arteries. Altogether, results point to phosphorylation of α_{1C} subunits at S1928 as a key biochemical event modulating vascular $Ca_V1.2$ activity and vascular function. These observations were exciting because they contrasted with experiments examining $Ca_V1.2$ channel regulation using cardiomyocytes as a general model cell that suggested no role of S1928 phosphorylation in PKA-dependent modulation of the channel (Ganesan, Maack, Johns, Sidor, & O'Rourke, 2006; Lemke et al., 2008; Roybal, Hennessey, & Marx, 2020). Instead, recent evidence points to the small Ras-like G protein Rad as a key effector mediating PKA-dependent regulation of cardiac $Ca_V1.2$ channels (Papa et al., 2021). Accordingly, Rad binding to cardiac $Ca_V1.2$ is suggested to have an inhibitory effect on channel function, which is removed upon PKA phosphorylation of Rad at several residues. Rad has also been suggested to modulate the activity of the predominant smooth muscle α_{1C} isoform when both proteins are over-expressed in Xenopus oocytes (Katz et al., 2021). However, a role for Rad in regulating $Ca_V1.2$ channels in native vascular smooth muscle remains to be established. Intriguingly, PKA-dependent regulation of neuronal $Ca_V1.2$ channels requires S1928 phosphorylation (Patriarchi et al., 2016; Qian et al., 2017). Thus, these results provide evidence of tissue-specific regulation of $Ca_V1.2$ channels, which requires further exploration (Man, Bartels, Horne, & Hell, 2020).

5.2 AC5

The activation of $Ca_V1.2$ in a PKA-dependent manner implies the production of the upstream second messenger cAMP to activate PKA and suggests the involvement of intermediate regulatory enzymes such as adenylate cyclase (AC) (Langeberg & Scott, 2015). In vascular smooth muscle cells from diabetic patients and animal models of this disease, alterations in AC signaling have been reported to contribute to cell proliferation, increased oxidative stress, impaired vasorelaxation, and enhanced expression of contractile genes (Hashim, Li, Nagakura, Takeo, & Anand-Srivastava, 2004; Hien et al., 2016; Li, Descorbeth, & Anand-Srivastava, 2008; Matsumoto, Wakabayashi, Kobayashi, & Kamata, 2005). Work by independent groups has linked the elevation in glucose levels with the production of the second messenger cAMP in yeast and murine pancreatic β cells (Dyachok et al., 2008; Munder & Kuntzel, 1989). This observation has been shared in vascular smooth muscle cells (Prada et al., 2019; Syed et al., 2019), where glucose triggered a modest albeit significant increase in cAMP synthesis as reported by a membrane-targeted cAMP biosensor (ICUE3) (Prada et al., 2019, 2020; Syed et al., 2019). The broad AC activator forskolin further amplified cAMP synthesis, hinting at a discrete activation of cAMP synthesis by high glucose (Syed et al., 2019). The question then was, what AC isoform was responsible for the generation of cAMP in response to elevations in extracellular glucose?

Nine membrane-bound AC isoforms have been described so far (Dessauer et al., 2017). Isoforms AC3, AC5, and AC6 are the most abundantly expressed in the vascular smooth muscle (Nelson et al., 2011; Ostrom et al., 2002; Webb, Yates, Yang, Mukhin, & Lanier, 2001). AC6 and AC3 are linked to β-adrenergic regulation of various K^+ channels that contribute to the relaxation of vascular smooth muscle (Nelson et al., 2011; Ostrom et al., 2002), whereas AC5 has been linked to enhanced contractility in diabetic hyperglycemia (Prada et al., 2019; Syed et al., 2019) (Fig. 2). Indeed, pharmacological inhibition and genetic ablation of AC5 prevented the glucose-induced cAMP production (Syed et al., 2019). This further correlated with a complete block of PKA-mediated potentiation of vascular $Ca_V1.2$ channels during hyperglycemia and diabetes (Syed et al., 2019). In addition to these findings, independent microarray analysis showed that AC5 expression was upregulated in vascular smooth muscle upon chronic hyperglycemic conditions (Hien et al., 2016). Altogether, these studies suggest an involvement of AC5 in the signaling cascade impacting vascular function during diabetic hyperglycemia, including potentiation of $Ca_V1.2$ activity leading to altered vascular smooth muscle contractility.

5.3 $P2Y_{11}$

AC activity is stimulated by signals from upstream G_S protein-coupled receptors (G_SPCR) (Halls & Cooper, 2017). A fundamental question was then, what G_SPCR could be engaged by diabetic hyperglycemia to induce AC5 activity? Purinergic (P2Y) receptors are GPCRs with ligand selectivity for nucleotides and nucleosides, with ATP as the primary physiological $P2Y_{11}$ agonist (Nishimura et al., 2017). The G protein subtype of each purinergic receptor defines the specificity of the intracellular signal elicited. $P2Y_{11}$ (human)/$P2Y_{11}$-like (murine) receptors have the unique property of coupling $G_{q/11}$ and G_s proteins (Kennedy, 2017). Therefore, $P2Y_{11}$ can stimulate AC/PKA and PLCβ/PKC signaling. Although $P2Y_{11}$ function in vascular smooth muscle cells is not entirely understood (Martin-Aragon Baudel, Espinosa-Tanguma, Nieves-Cintron, & Navedo, 2020), a polymorphism (Ala-87-Thr) in $P2Y_{11}$ has been associated with an increased risk of acute myocardial infarction and C-reactive protein blood levels (Amisten, Melander, Wihlborg, Berglund, & Erlinge, 2007). In cardiac cells, $P2Y_{11}$ has been shown to mediate positive ionotropic effects via PLC and cAMP signaling activation (Balogh et al., 2005). Its function seems impaired in a desmin-deficient mouse model of cardiomyopathy, leading to congestive heart failure (Balogh et al., 2005). In colonic smooth muscle, $P2Y_{11}$ receptors are involved in fast and slow relaxations through K_{Ca} channels, conforming to parasympathetic inhibition of the gut (King & Townsend-Nicholson, 2008). In vascular smooth muscle cells from rat pulmonary arteries, a Ca^{2+}-dependent chloride current was activated by ATP involving a P2Y receptor that was suggested to resemble a $P2Y_{11}$ receptor (Chootip, Gurney, & Kennedy, 2005).

A recent study linked $P2Y_{11}$ to glucose-mediated regulation of vascular smooth muscle excitability (Prada et al., 2019). Interest in $P2Y_{11}$ in this axis stems from the discovery that elevated glucose could stimulate vascular $Ca_V1.2$ channels activity through a mechanism requiring PKA, leading to increased intracellular Ca^{2+} and vasoconstriction of mouse cerebral arteries (Nystoriak et al., 2017). $P2Y_{11}$ is the only purinergic receptor able to couple G_s activity and, therefore, AC/PKA, so it represents an ideal candidate to mediate this mechanism. In human and mouse vascular smooth muscle cells expressing Förster resonance energy transfer (FRET) based cAMP biosensors, the highly selective $P2Y_{11}$ agonist NF546 increased the cAMP synthesis (Prada et al., 2019). This NF546-induced cAMP response was blocked in cells treated with the selective $P2Y_{11}$ inhibitor NF340 but not with selective $P2Y_1$ and $P2Y_6$ inhibitors. Notably, NF546 elicited a cAMP synthesis similar in magnitude to the response triggered by 20 mM D-glucose. Simultaneous application of NF546/D-glucose produced a cAMP synthesis equivalent to that of NF546 or D-glucose alone, suggesting that they may act through the same signaling pathway. Moreover, glucose and NF546-induced cAMP synthesis (as an independent stimulus or combination) in human and mouse vascular smooth muscle cells was blocked if cells were first pre-treated with NF340. These findings proved the involvement of human $P2Y_{11}$ and mouse $P2Y_{11}$-like receptors in hyperglycemia-induced cAMP synthesis. Moreover, data implicated $P2Y_{11}$/$P2Y_{11}$-like receptors in the mechanism by which diabetic hyperglycemia alters $Ca_V1.2$ channel activity and vascular reactivity (Fig. 2). Thus, $P2Y_{11}$/$P2Y_{11}$-like receptors may be a new therapeutic target for vascular complications during diabetic hyperglycemia.

5.4 A-kinase anchoring proteins (AKAPs)

With the identification of signal generators and effector proteins involved in modulating $Ca_V1.2$ channel activity and vascular reactivity during diabetic hyperglycemia, a remaining issue was how $P2Y_{11}$, AC5, PKA, and $Ca_V1.2$

could effectively communicate with each other. AKAPs have been critical in fine-tuning signal transduction pathways by positioning signaling proteins close to specific substrates (Langeberg & Scott, 2015; Man, Navedo, Horne, & Hell, 2019). AKAP5 (murine AKAP150 and human AKAP79) has been shown to mediate the rate and spatial precision of $Ca_V1.2$ phosphorylation and dephosphorylation events in different cellular types (Dai, Hall, & Hell, 2009; Li, Nooh, & Bahouth, 2013). Accordingly, AKAP5 has been shown to anchor the PKA-dependent phosphorylation and enhance $Ca_V1.2$ activity and calcineurin, a phosphatase that depresses the $Ca_V1.2$ channel in the neuronal dendritic spines activity (Sanderson & Dell'Acqua, 2011). AKAP5 has been shown to facilitate the regulation of Ca^{2+} sparklet activity in vascular smooth muscle by scaffolding protein kinase and phosphatase near the $Ca_V1.2$ subunit (Navedo et al., 2008). A role for AKAP5-anchored $Ca_V1.2$ nanodomains in regulating vascular function was demonstrated in arteries from AKAP5 knockout mice. Arteries from these animals had lower $[Ca^{2+}]_i$ and developed less myogenic tone than wild-type (WT) arteries. Moreover, AKAP5 knockout mice displayed a hypotensive phenotype and did not develop hypertension when chronically infused with the vasoconstrictor angiotensin II (Navedo et al., 2008). These results provide a clear pathophysiological role for AKAP5-anchored $Ca_V1.2$ nanodomains in regulating vascular function.

A mechanistic framework involving AKAP5 in regulating $Ca_V1.2$ channels during diabetic hyperglycemia was recently put forward (Fig. 2). Multipronged approaches provided evidence for the role of AKAP5 in supporting a nanomolecular complex involving $P2Y_{11}$, AC5, PKA, and $Ca_V1.2$ in diabetes (Prada et al., 2020). Super-resolution localization maps showed that $P2Y_{11}$, AC5, PKA, and $Ca_V1.2$ are part of a macromolecular complex orchestrated by AKAP5 under hyperglycemic conditions and in murine models of diabetes. The formation of this macromolecular complex correlates functionally with enhanced S1928 phosphorylation and $Ca_V1.2$ activity in response to hyperglycemia and diabetes. In AKAP5 KO mice, the $P2Y_{11}$/AC5/PKA/$Ca_V1.2$ association is reduced, which prevented the stimulation of $Ca_V1.2$ channels by hyperglycemic conditions in cerebral smooth muscle and arteries from diabetic mice (Nystoriak et al., 2017; Prada et al., 2020).

Furthermore, the AKAP5-driven nanocomplex is essential for enhanced vasoconstriction and changes in blood flow during diabetic hyperglycemia (Prada et al., 2020). Disruption of the interaction between AKAP5 and

PKA, with genetically modified mice expressing an AKAP5 that could not bind PKA or PKC confirmed an essential role for compartmentalized PKA, but not compartmentalized PKC, in mediating the glucose effects on $Ca_V1.2$ channel activity and cerebral vascular reactivity (Nystoriak et al., 2017; Prada et al., 2020). These exciting results provided a mechanism by which hyperglycemia/diabetes activates nanostructural domains mediated by AKAP5 to recruit $P2Y_{11}$, AC5, and PKA, leading to phosphorylation of $Ca_V1.2$ at S1928, potentiation of $Ca_V1.2$ activity, vasoconstriction and alterations in blood flow and blood pressure.

6. Conclusions and future directions

The L-type Ca^{2+} channel $Ca_V1.2$ plays a crucial role in regulating intracellular Ca^{2+}, gene expression, and vascular smooth muscle contraction. Thus, it is unsurprising that altered $Ca_V1.2$ channel function contributes to vascular complications associated with numerous pathological conditions such as diabetes. Here, we discussed the general properties of $Ca_V1.2$ structure and its regulation, including how $Ca_V1.2$ channel activity may be influenced by α_{1C} splice variants and phosphorylation events mediated by different kinases. This narrative paved the way to examine and describe the current understanding of how $Ca_V1.2$ channels in vascular smooth muscle are regulated by elevations in extracellular glucose and during diabetes. Intriguingly, the in-depth examination of how diabetes alters vascular $Ca_V1.2$ channels revealed a new nanomolecular complex (AKAP5/$P2Y_{11}$/AC5/PKA/$Ca_V1.2$) with profound implications for the regulation of vascular reactivity.

Moreover, this nanocomplex can have a broad systemic impact as most, if not all, members of the complex are broadly expressed in other cells. There are also some areas of opportunity for future research. For example, it will be important to determine where the nucleotide signal that activates $P2Y_{11}$ receptors in vascular smooth muscle cells in response to diabetic hyperglycemia comes from. What is the nature of this nucleotide signal—e.g., ATP? Is this nucleotide signal dependent on an autocrine, paracrine, or combined mechanism? Another area of interest will be to determine if the dynamics of the AKAP5/$P2Y_{11}$/AC5/PKA/$Ca_V1.2$ nanodomain are impacted by hyperglycemia and diabetes and whether such change can be prevented by blocking any component of the pathway or currently unclear upstream signals.

Acknowledgments

This work was supported by American Heart Association grants 852984 (M.N.C.), 830626 (M.MA.B), and R01HL149127 and 161872 (M.F.N).

References

Abd-Elrahman, K. S., Colinas, O., Walsh, E. J., Zhu, H. L., Campbell, C. M., Walsh, M. P., et al. (2017). Abnormal myosin phosphatase targeting subunit 1 phosphorylation and actin polymerization contribute to impaired myogenic regulation of cerebral arterial diameter in the type 2 diabetic Goto-Kakizaki rat. *Journal of Cerebral Blood Flow and Metabolism*, *37*(1), 227–240. https://doi.org/10.1177/0271678X15622463.

Aiello, E. A., Walsh, M. P., & Cole, W. C. (1995). Phosphorylation by protein kinase A enhances delayed rectifier K+ current in rabbit vascular smooth muscle cells. *The American Journal of Physiology*, *268*(2 Pt. 2), H926–H934. http://www.ncbi.nlm.nih.gov/entrez/query.fcgi?cmd=Retrieve&db=PubMed&dopt=Citation&list_uids=7864221. http://ajpheart.physiology.org/content/268/2/H926.full.pdf.

Amberg, G. C., & Navedo, M. F. (2013). Calcium dynamics in vascular smooth muscle. *Microcirculation*, *20*(4), 281–289. https://doi.org/10.1111/micc.12046.

Amisten, S., Melander, O., Wihlborg, A. K., Berglund, G., & Erlinge, D. (2007). Increased risk of acute myocardial infarction and elevated levels of C-reactive protein in carriers of the Thr-87 variant of the ATP receptor P2Y11. *European Heart Journal*, *28*(1), 13–18. https://doi.org/10.1093/eurheartj/ehl410.

Arikkath, J., & Campbell, K. P. (2003). Auxiliary subunits: Essential components of the voltage-gated calcium channel complex. *Current Opinion in Neurobiology*, *13*(3), 298–307. https://doi.org/10.1016/s0959-4388(03)00066-7.

Balogh, J., Wihlborg, A. K., Isackson, H., Joshi, B. V., Jacobson, K. A., Arner, A., et al. (2005). Phospholipase C and cAMP-dependent positive inotropic effects of ATP in mouse cardiomyocytes via P2Y11-like receptors. *Journal of Molecular and Cellular Cardiology*, *39*(2), 223–230. https://doi.org/10.1016/j.yjmcc.2005.03.007.

Bannister, J. P., Adebiyi, A., Zhao, G., Narayanan, D., Thomas, C. M., Feng, J. Y., et al. (2009). Smooth muscle cell alpha2delta-1 subunits are essential for vasoregulation by CaV1.2 channels [Research Support, N.I.H., Extramural Research Support, Non-U.S. Gov't]. *Circulation Research*, *105*(10), 948–955. https://doi.org/10.1161/CIRCRESAHA.109.203620.

Bannister, J. P., Bulley, S., Narayanan, D., Thomas-Gatewood, C., Luzny, P., Pachuau, J., et al. (2012). Transcriptional upregulation of alpha2delta-1 elevates arterial smooth muscle cell voltage-dependent Ca2+ channel surface expression and cerebrovascular constriction in genetic hypertension. *Hypertension*, *60*(4), 1006–1015. https://doi.org/10.1161/HYPERTENSIONAHA.112.199661.

Bannister, J. P., Thomas-Gatewood, C. M., Neeb, Z. P., Adebiyi, A., Cheng, X., & Jaggar, J. H. (2011). CaV1.2 channel N-terminal splice variants modulate functional surface expression in resistance size artery smooth muscle cells. *Journal of Biological Chemistry*, *286*(17), 15058–15066. https://doi.org/10.1074/jbc.m110.182816.

Blatter, L. A., & Wier, W. G. (1994). Nitric oxide decreases [Ca2+]i in vascular smooth muscle by inhibition of the calcium current. *Cell Calcium*, *15*(2), 122–131. http://www.ncbi.nlm.nih.gov/htbin-post/Entrez/query?db=m&form=6&dopt=r&uid=8149412.

Burgess, D. L., Gefrides, L. A., Foreman, P. J., & Noebels, J. L. (2001). A cluster of three novel Ca2+ channel gamma subunit genes on chromosome 19q13.4: Evolution and expression profile of the gamma subunit gene family. *Genomics*, *71*(3), 339–350. https://doi.org/10.1006/geno.2000.6440.

Caldwell, B. V. (1993). Treating hypertension in the diabetic patient: Therapeutic goals and the role of calcium channel blockers. *Clinical Therapeutics*, *15*(4), 618–636. Discussion 617. https://www.ncbi.nlm.nih.gov/pubmed/8221813.

Carmines, P. K., Ohishi, K., & Ikenaga, H. (1996). Functional impairment of renal afferent arteriolar voltage-gated calcium channels in rats with diabetes mellitus. *The Journal of Clinical Investigation*, *98*(11), 2564–2571. https://doi.org/10.1172/JCI119075.

Carvajal, J. A., Germain, A. M., Huidobro-Toro, J. P., & Weiner, C. P. (2000). Molecular mechanism of cGMP-mediated smooth muscle relaxation. *Journal of Cellular Physiology*, *184*(3), 409–420. https://doi.org/10.1002/1097-4652(200009)184:3<409::AID-JCP16>3.0.CO;2-K.

Catterall, W. A. (2000). Structure and regulation of voltage-gated Ca^{2+} channels. *Annual Review of Cell and Developmental Biology*, *16*, 521–555. http://www.ncbi.nlm.nih.gov/htbin-post/Entrez/query?db=m&form=6&dopt=r&uid=11031246. http://cellbio.annualreviews.org/cgi/content/full/16/1/521. http://cellbio.annualreviews.org/cgi/content/abstract/16/1/521.

Catterall, W. A. (2011). Voltage-gated calcium channels. *Cold Spring Harbor Perspectives in Biology*, *3*(8), a003947. https://doi.org/10.1101/cshperspect.a003947.

Cheng, X., Liu, J., Asuncion-Chin, M., Blaskova, E., Bannister, J. P., Dopico, A. M., et al. (2007). A novel Ca(V)1.2 N terminus expressed in smooth muscle cells of resistance size arteries modifies channel regulation by auxiliary subunits. *The Journal of Biological Chemistry*, *282*(40), 29211–29221. https://doi.org/10.1074/jbc.M610623200.

Cheng, X., Pachuau, J., Blaskova, E., Asuncion-Chin, M., Liu, J., Dopico, A. M., et al. (2009). Alternative splicing of Cav1.2 channel exons in smooth muscle cells of resistance-size arteries generates currents with unique electrophysiological properties. *American Journal of Physiology. Heart and Circulatory Physiology*, *297*(2), H680–H688. https://doi.org/10.1152/ajpheart.00109.2009.

Chootip, K., Gurney, A. M., & Kennedy, C. (2005). Multiple P2Y receptors couple to calcium-dependent, chloride channels in smooth muscle cells of the rat pulmonary artery. *Respiratory Research*, *6*(1), 124. https://doi.org/10.1186/1465-9921-6-124.

Cobine, C. A., Callaghan, B. P., & Keef, K. D. (2007). Role of L-type calcium channels and PKC in active tone development in rabbit coronary artery. *American Journal of Physiology. Heart and Circulatory Physiology*, *292*(6), H3079–88. https://doi.org/10.1152/ajpheart.01261.2006. http://www.ncbi.nlm.nih.gov/entrez/query.fcgi?cmd=Retrieve&db=PubMed&dopt=Citation&list_uids=17277031.

Cox, R. H., & Fromme, S. (2015). Expression of calcium channel subunit variants in small mesenteric arteries of WKY and SHR. *American Journal of Hypertension*, *28*(10), 1229–1239. https://doi.org/10.1093/ajh/hpv024.

Dai, S., Hall, D. D., & Hell, J. W. (2009). Supramolecular assemblies and localized regulation of voltage-gated ion channels. *Physiological Reviews*, *89*(2), 411–452. https://doi.org/10.1152/physrev.00029.2007.

Davis, M. J., & Hill, M. A. (1999). Signaling mechanisms underlying the vascular myogenic response. *Physiological Reviews*, *79*(2), 387–423. http://www.ncbi.nlm.nih.gov/pubmed/10221985.

De Jongh, K. S., Murphy, B. J., Colvin, A. A., Hell, J. W., Takahashi, M., & Catterall, W. A. (1996). Specific phosphorylation of a site in the full-length form of the alpha 1 subunit of the cardiac L-type calcium channel by adenosine 3',5'-cyclic monophosphate-dependent protein kinase. *Biochemistry*, *35*(32), 10392–10402. http://www.ncbi.nlm.nih.gov/htbin-post/Entrez/query?db=m&form=6&dopt=r&uid=8756695. https://doi.org/10.1021/bi953023c.

Dessauer, C. W., Watts, V. J., Ostrom, R. S., Conti, M., Dove, S., & Seifert, R. (2017). International Union of Basic and Clinical Pharmacology. CI. Structures and small molecule modulators of mammalian adenylyl cyclases. *Pharmacological Reviews*, *69*(2), 93–139. https://doi.org/10.1124/pr.116.013078.

Diebold, R. J., Koch, W. J., Ellinor, P. T., Wang, J. J., Muthuchamy, M., Wieczorek, D. F., et al. (1992). Mutually exclusive exon splicing of the cardiac calcium channel alpha 1 subunit gene generates developmentally regulated isoforms in the rat heart. *Proceedings of the National Academy of Sciences of the United States of America*, *89*(4), 1497–1501. https://doi.org/10.1073/pnas.89.4.1497.

Dolphin, A. C. (2003). Beta subunits of voltage-gated calcium channels. *Journal of Bioenergetics and Biomembranes*, *35*(6), 599–620. https://doi.org/10.1023/b:jobb.0000008026.37790.5a.

Dolphin, A. C. (2016). Voltage-gated calcium channels and their auxiliary subunits: Physiology and pathophysiology and pharmacology. *The Journal of Physiology*, *594*(19), 5369–5390. https://doi.org/10.1113/JP272262.

Dyachok, O., Idevall-Hagren, O., Sagetorp, J., Tian, G., Wuttke, A., Arrieumerlou, C., et al. (2008). Glucose-induced cyclic AMP oscillations regulate pulsatile insulin secretion. *Cell Metabolism*, *8*(1), 26–37. https://doi.org/10.1016/j.cmet.2008.06.003.

Fan, J., Fan, W., Lei, J., Zhou, Y., Xu, H., Kapoor, I., et al. (2019). Galectin-1 attenuates cardiomyocyte hypertrophy through splice-variant specific modulation of CaV1.2 calcium channel. *Biochimica et Biophysica Acta—Molecular Basis of Disease*, *1865*(1), 218–229. https://doi.org/10.1016/j.bbadis.2018.08.016.

Fluitt, M. B., Mohit, N., Gambhir, K. K., & Nunlee-Bland, G. (2022). To the future: The role of exosome-derived microRNAs as markers, mediators, and therapies for endothelial dysfunction in type 2 diabetes mellitus. *Journal Diabetes Research*, *2022*, 5126968. https://doi.org/10.1155/2022/5126968.

Francis, S. H., Noblett, B. D., Todd, B. W., Wells, J. N., & Corbin, J. D. (1988). Relaxation of vascular and tracheal smooth muscle by cyclic nucleotide analogs that preferentially activate purified cGMP-dependent protein kinase. *Molecular Pharmacology*, *34*(4), 506–517. https://www.ncbi.nlm.nih.gov/pubmed/2845250.

Gallegos, L. L., Kunkel, M. T., & Newton, A. C. (2006). Targeting protein kinase C activity reporter to discrete intracellular regions reveals spatiotemporal differences in agonist-dependent signaling. *The Journal of Biological Chemistry*, *281*(41), 30947–30956. https://doi.org/10.1074/jbc.M603741200.

Ganesan, A. N., Maack, C., Johns, D. C., Sidor, A., & O'Rourke, B. (2006). Beta-adrenergic stimulation of L-type Ca2+ channels in cardiac myocytes requires the distal carboxyl terminus of alpha1C but not serine 1928. *Circulation Research*, *98*(2), e11–e18. https://doi.org/10.1161/01.RES.0000202692.23001.e2.

Ghosh, D., Syed, A. U., Prada, M. P., Nystoriak, M. A., Santana, L. F., Nieves-Cintron, M., et al. (2017). Calcium channels in vascular smooth muscle. *Advances in Pharmacology*, *78*, 49–87. https://doi.org/10.1016/bs.apha.2016.08.002.

Gidh-Jain, M., Huang, B., Jain, P., Battula, V., & el-Sherif, N. (1995). Reemergence of the fetal pattern of L-type calcium channel gene expression in non infarcted myocardium during left ventricular remodeling. *Biochemical and Biophysical Research Communications*, *216*(3), 892–897. http://www.ncbi.nlm.nih.gov/entrez/query.fcgi?cmd=Retrieve&db=PubMed&dopt=Citation&list_uids=7488209.

Gurnett, C. A., De Waard, M., & Campbell, K. P. (1996). Dual function of the voltage-dependent Ca2+ channel alpha 2 delta subunit in current stimulation and subunit interaction. *Neuron*, *16*(2), 431–440. http://www.ncbi.nlm.nih.gov/pubmed/8789958. http://ac.els-cdn.com/S0896627300800616/1-s2.0-S0896627300800616-main.pdf?_tid=e892109e-e428-11e5-933e-00000aacb360&acdnat=1457330185_467eb9866f6a2435ad4ebb4772eaeb15.

Halls, M. L., & Cooper, D. M. (2011). Regulation by Ca2+-signaling pathways of adenylyl cyclases. *Cold Spring Harbor Perspectives in Biology*, *3*(1), a004143. https://doi.org/10.1101/cshperspect.a004143.

Halls, M. L., & Cooper, D. M. F. (2017). Adenylyl cyclase signalling complexes—Pharmacological challenges and opportunities. *Pharmacology & Therapeutics*, *172*, 171–180. https://doi.org/10.1016/j.pharmthera.2017.01.001.

Hashim, S., Li, Y., & Anand-Srivastava, M. B. (2006). G protein-linked cell signaling and cardiovascular functions in diabetes/hyperglycemia. *Cell Biochemistry and Biophysics*, *44*(1), 51–64. CBB:44:1:051 [pii] 10.1385/CBB:44:1:051.
Hashim, S., Li, Y., Nagakura, A., Takeo, S., & Anand-Srivastava, M. B. (2004). Modulation of G-protein expression and adenylyl cyclase signaling by high glucose in vascular smooth muscle. *Cardiovascular Research*, *63*(4), 709–718. https://doi.org/10.1016/j.cardiores.2004.04.021.
Hien, T. T., Turczynska, K. M., Dahan, D., Ekman, M., Grossi, M., Sjogren, J., et al. (2016). Elevated glucose levels promote contractile and cytoskeletal gene expression in vascular smooth muscle via rho/protein kinase C and actin polymerization. *The Journal of Biological Chemistry*, *291*(7), 3552–3568. https://doi.org/10.1074/jbc.M115.654384.
Houslay, M. D. (2010). Underpinning compartmentalised cAMP signalling through targeted cAMP breakdown. *Trends in Biochemical Sciences*, *35*(2), 91–100. https://doi.org/10.1016/j.tibs.2009.09.007.
Hu, Z., Liang, M., & Soong, T. (2017). Alternative splicing of L-type CaV1.2 calcium channels: Implications in cardiovascular diseases. *Genes*, *8*(12), 344. https://doi.org/10.3390/genes8120344.
Hu, H., & Marban, E. (1998). Isoform-specific inhibition of L-type calcium channels by dihydropyridines is independent of isoform-specific gating properties. *Molecular Pharmacology*, *53*(5), 902–907. http://www.ncbi.nlm.nih.gov/htbin-post/Entrez/query?db=m&form=6&dopt=r&uid=9584217. http://www.molpharm.org/cgi/content/full/53/5/902.
Ishikawa, T., Hume, J. R., & Keef, K. D. (1993). Regulation of Ca2+ channels by cAMP and cGMP in vascular smooth muscle cells. *Circulation Research*, *73*(6), 1128–1137. http://www.ncbi.nlm.nih.gov/entrez/query.fcgi?cmd=Retrieve&db=PubMed&dopt=Citation&list_uids=8222084.
Katz, M., Subramaniam, S., Chomsky-Hecht, O., Tsemakhovich, V., Flockerzi, V., Klussmann, E., et al. (2021). Reconstitution of beta-adrenergic regulation of CaV1.2: Rad-dependent and Rad-independent protein kinase A mechanisms. *Proceedings of the National Academy of Sciences of the United States of America*, *118*(21), e2100021118. https://doi.org/10.1073/pnas.2100021118.
Keef, K. D., Hume, J. R., & Zhong, J. (2001). Regulation of cardiac and smooth muscle Ca(2+) channels (Ca(V)1.2a,b) by protein kinases. *American Journal of Physiology. Cell Physiology*, *281*(6), C1743–C1756. http://www.ncbi.nlm.nih.gov/pubmed/11698232.
Kennedy, C. (2017). P2Y11 Receptors: Properties, distribution and functions. *Advances in Experimental Medicine and Biology*, *1051*, 107–122. https://doi.org/10.1007/5584_2017_89.
Kharade, S. V., Sonkusare, S. K., Srivastava, A. K., Thakali, K. M., Fletcher, T. W., Rhee, S. W., et al. (2013). The beta3 subunit contributes to vascular calcium channel upregulation and hypertension in angiotensin II-infused C57BL/6 mice. *Hypertension*, *61*(1), 137–142. https://doi.org/10.1161/HYPERTENSIONAHA.112.197863.
King, B. F., & Townsend-Nicholson, A. (2008). Involvement of P2Y1 and P2Y11 purinoceptors in parasympathetic inhibition of colonic smooth muscle. *The Journal of Pharmacology and Experimental Therapeutics*, *324*(3), 1055–1063. https://doi.org/10.1124/jpet.107.131169.
Knot, H. J., & Nelson, M. T. (1998). Regulation of arterial diameter and wall [Ca^{2+}] in cerebral arteries of rat by membrane potential and intravascular pressure. *The Journal of Physiology*, *508*(Pt. 1), 199–209. http://www.ncbi.nlm.nih.gov/htbin-post/Entrez/query?db=m&form=6&dopt=r&uid=9490839. http://www.journals.cup.org/owa_dba/owa/approval?sjid=PHY&said=7258&spii=S0022375198072585.
Kold-Petersen, H., Brondum, E., Nilsson, H., Flyvbjerg, A., & Aalkjaer, C. (2012). Impaired myogenic tone in isolated cerebral and coronary resistance arteries from the goto-kakizaki rat model of type 2 diabetes. *Journal of Vascular Research*, *49*(3), 267–278. https://doi.org/10.1159/000335487.

Langeberg, L. K., & Scott, J. D. (2005). A-kinase-anchoring proteins. *Journal of Cell Science, 118*(Pt. 15), 3217–3220. https://doi.org/10.1242/jcs.02416.

Langeberg, L. K., & Scott, J. D. (2015). Signalling scaffolds and local organization of cellular behaviour. *Nature Reviews. Molecular Cell Biology, 16*(4), 232–244. https://doi.org/10.1038/nrm3966.

Lemke, T., Welling, A., Christel, C. J., Blaich, A., Bernhard, D., Lenhardt, P., et al. (2008). Unchanged beta-adrenergic stimulation of cardiac L-type calcium channels in Ca v 1.2 phosphorylation site S1928A mutant mice. *The Journal of Biological Chemistry, 283*(50), 34738–34744. https://doi.org/10.1074/jbc.M804981200.

Li, Y., Descorbeth, M., & Anand-Srivastava, M. B. (2008). Role of oxidative stress in high glucose-induced decreased expression of Gialpha proteins and adenylyl cyclase signaling in vascular smooth muscle cells. *American Journal of Physiology. Heart and Circulatory Physiology, 294*(6), H2845–H2854. https://doi.org/10.1152/ajpheart.91422.2007.

Li, X., Nooh, M. M., & Bahouth, S. W. (2013). Role of AKAP79/150 protein in β1-adrenergic receptor trafficking and signaling in mammalian cells. *The Journal of Biological Chemistry, 288*(47), 33797–33812. https://doi.org/10.1074/jbc.M113.470559.

Liao, P., Li, G., Yu, D. J., Yong, T. F., Wang, J. J., Wang, J., et al. (2009). Molecular alteration of Ca(v)1.2 calcium channel in chronic myocardial infarction. *Pflügers Archiv, 458*(4), 701–711. https://doi.org/10.1007/s00424-009-0652-4.

Liao, P., & Soong, T. W. (2010). Understanding alternative splicing of Cav1.2 calcium channels for a new approach towards individualized medicine. *Journal of Biomedical Research, 24*(3), 181–186. https://doi.org/10.1016/S1674-8301(10)60027-9.

Liao, P., Yong, T. F., Liang, M. C., Yue, D. T., & Soong, T. W. (2005). Splicing for alternative structures of Cav1.2 Ca2+ channels in cardiac and smooth muscles. *Cardiovascular Research, 68*(2), 197–203. https://doi.org/10.1016/j.cardiores.2005.06.024.

Liao, P., Yu, D., Li, G., Yong, T. F., Soon, J. L., Chua, Y. L., et al. (2007). A smooth muscle Cav1.2 calcium channel splice variant underlies hyperpolarized window current and enhanced state-dependent inhibition by nifedipine. *Journal of Biological Chemistry, 282*(48), 35133–35142. https://doi.org/10.1074/jbc.m705478200.

Liao, P., Yu, D., Lu, S., Tang, Z., Liang, M. C., Zeng, S., et al. (2004). Smooth muscle-selective alternatively spliced exon generates functional variation in Cav1.2 calcium channels. *The Journal of Biological Chemistry, 279*(48), 50329–50335. https://doi.org/10.1074/jbc.M409436200.

Ma, Y. G., Wang, J. W., Bai, Y. G., Liu, M., Xie, M. J., & Dai, Z. J. (2017). Salidroside contributes to reducing blood pressure and alleviating cerebrovascular contractile activity in diabetic Goto-Kakizaki Rats by inhibition of L-type calcium channel in smooth muscle cells. *BMC Pharmacology & Toxicology, 18*(1), 30. https://doi.org/10.1186/s40360-017-0135-8.

Man, K. N. M., Bartels, P., Horne, M. C., & Hell, J. W. (2020). Tissue-specific adrenergic regulation of the L-type Ca^{2+} channel CaV1.2. *Science Signaling, 13*(663), eabc6438. https://doi.org/10.1126/scisignal.abc6438.

Man, K. N. M., Navedo, M. F., Horne, M. C., & Hell, J. W. (2019). Beta2 adrenergic receptor complexes with the L-type Ca^{2+} channel CaV1.2 and AMPA-type glutamate receptors: Paradigms for pharmacological targeting of protein interactions. *Annual Review of Pharmacology and Toxicology, 6*(60), 155–174. https://doi.org/10.1146/annurev-pharmtox-010919-023404.

Martin-Aragon Baudel, M., Espinosa-Tanguma, R., Nieves-Cintron, M., & Navedo, M. F. (2020). Purinergic signaling during hyperglycemia in vascular smooth muscle cells [review]. *Frontiers in Endocrinology, 11*, 329. https://doi.org/10.3389/fendo.2020.00329.

Matsumoto, T., Wakabayashi, K., Kobayashi, T., & Kamata, K. (2005). Functional changes in adenylyl cyclases and associated decreases in relaxation responses in mesenteric arteries from diabetic rats. *American Journal of Physiology. Heart and Circulatory Physiology, 289*(5), H2234–H2243. https://doi.org/10.1152/ajpheart.00971.2004.

Moore, C. L., Nelson, P. L., Parelkar, N. K., Rusch, N. J., & Rhee, S. W. (2014). Protein kinase A-phosphorylated KV1 channels in PSD95 signaling complex contribute to the resting membrane potential and diameter of cerebral arteries. *Circulation Research, 114*(8), 1258–1267. https://doi.org/10.1161/CIRCRESAHA.114.303167.

Moosmang, S., Schulla, V., Welling, A., Feil, R., Feil, S., Wegener, J. W., et al. (2003). Dominant role of smooth muscle L-type calcium channel Cav1.2 for blood pressure regulation. *The EMBO Journal, 22*(22), 6027–6034. http://www.ncbi.nlm.nih.gov/entrez/query.fcgi?cmd=Retrieve&db=PubMed&dopt=Citation&list_uids=14609949.

Munder, T., & Kuntzel, H. (1989). Glucose-induced cAMP signaling in Saccharomyces cerevisiae is mediated by the CDC25 protein. *FEBS Letters, 242*(2), 341–345. https://doi.org/10.1016/0014-5793(89)80498-3.

Navedo, M. F., & Amberg, G. C. (2013). Local regulation of L-type ca(2+) channel sparklets in arterial smooth muscle. *Microcirculation, 20*(4), 290–298. https://doi.org/10.1111/micc.12021.

Navedo, M. F., Amberg, G. C., Nieves, M., Molkentin, J. D., & Santana, L. F. (2006). Mechanisms underlying heterogeneous Ca^{2+} sparklet activity in arterial smooth muscle. *Journal of General Physiology, 127*(6), 611–622.

Navedo, M. F., Amberg, G., Votaw, S. V., & Santana, L. F. (2005). Constitutively active L-type Ca^{2+} channels. *Proceedings of the National Academy of Sciences of the United States of America, 102*(31), 11112–11117.

Navedo, M. F., Cheng, E. P., Yuan, C., Votaw, S., Molkentin, J. D., Scott, J. D., et al. (2010). Increased coupled gating of L-type Ca2+ channels during hypertension and Timothy syndrome [Research Support, N.I.H., Extramural Research Support, Non-U.S. Gov't]. *Circulation Research, 106*(4), 748–756. https://doi.org/10.1161/CIRCRESAHA.109.213363.

Navedo, M. F., Nieves-Cintron, M., Amberg, G. C., Yuan, C., Votaw, V. S., Lederer, W. J., et al. (2008). AKAP150 is required for stuttering persistent Ca2+ sparklets and angiotensin II-induced hypertension. *Circulation Research, 102*(2), e1–e11. doi: CIRCRESAHA.107.167809 [pii] https://doi.org/10.1161/CIRCRESAHA.107.167809.

Navedo, M. F., & Santana, L. F. (2013). CaV1.2 sparklets in heart and vascular smooth muscle. *Journal of Molecular and Cellular Cardiology, 58*, 67–76. https://doi.org/10.1016/j.yjmcc.2012.11.018.

Navedo, M. F., Takeda, Y., Nieves-Cintron, M., Molkentin, J. D., & Santana, L. F. (2010). Elevated Ca2+ sparklet activity during acute hyperglycemia and diabetes in cerebral arterial smooth muscle cells [Research Support, N.I.H., Extramural Research Support, Non-U.S. Gov't]. *American Journal of Physiology—Cell Physiology, 298*(2), C211–C220. https://doi.org/10.1152/ajpcell.00267.2009.

Nelson, C. P., Rainbow, R. D., Brignell, J. L., Perry, M. D., Willets, J. M., Davies, N. W., et al. (2011). Principal role of adenylyl cyclase 6 in K(+) channel regulation and vasodilator signalling in vascular smooth muscle cells. *Cardiovascular Research, 91*(4), 694–702. https://doi.org/10.1093/cvr/cvr137.

Nieves-Cintron, M., Amberg, G. C., Navedo, M. F., Molkentin, J. D., & Santana, L. F. (2008). The control of Ca2+ influx and NFATc3 signaling in arterial smooth muscle during hypertension. *Proceedings of the National Academy of Sciences of the United States of America, 105*(40), 15623–15628. doi: 0808759105 [pii] https://doi.org/10.1073/pnas.0808759105.

Nieves-Cintron, M., Flores-Tamez, V. A., Le, T., Baudel, M. M., & Navedo, M. F. (2021). Cellular and molecular effects of hyperglycemia on ion channels in vascular smooth muscle. *Cellular and Molecular Life Sciences, 78*(1), 31–61. https://doi.org/10.1007/s00018-020-03582-z.

Nishimura, A., Sunggip, C., Oda, S., Numaga-Tomita, T., Tsuda, M., & Nishida, M. (2017). Purinergic P2Y receptors: Molecular diversity and implications for treatment of cardiovascular diseases. *Pharmacology & Therapeutics*, *180*, 113–128. https://doi.org/10.1016/j.pharmthera.2017.06.010.

Nystoriak, M. A., Murakami, K., Penar, P. L., & Wellman, G. C. (2009). Ca(v)1.2 splice variant with exon 9* is critical for regulation of cerebral artery diameter. *American Journal of Physiology. Heart and Circulatory Physiology*, *297*(5), H1820–H1828. https://doi.org/10.1152/ajpheart.00326.2009.

Nystoriak, M. A., Nieves-Cintron, M., Patriarchi, T., Buonarati, O. R., Prada, M. P., Morotti, S., et al. (2017). Ser1928 phosphorylation by PKA stimulates the L-type Ca2+ channel CaV1.2 and vasoconstriction during acute hyperglycemia and diabetes. *Science Signaling*, *10*(463), eaaf9647. https://doi.org/10.1126/scisignal.aaf9647.

Ostrom, R. S., Liu, X., Head, B. P., Gregorian, C., Seasholtz, T. M., & Insel, P. A. (2002). Localization of adenylyl cyclase isoforms and G protein-coupled receptors in vascular smooth muscle cells: Expression in caveolin-rich and noncaveolin domains. *Molecular Pharmacology*, *62*(5), 983–992. http://www.ncbi.nlm.nih.gov/pubmed/12391260.

Papa, A., Kushner, J., Hennessey, J. A., Katchman, A. N., Zakharov, S. I., Chen, B. X., et al. (2021). Adrenergic CaV1.2 activation via rad phosphorylation converges at alpha1C I-II loop. *Circulation Research*, *128*(1), 76–88. https://doi.org/10.1161/CIRCRESAHA.120.317839.

Patel, N. A., Chalfant, C. E., Yamamoto, M., Watson, J. E., Eichler, D. C., & Cooper, D. R. (1999). Acute hyperglycemia regulates transcription and posttranscriptional stability of PKCbetaII mRNA in vascular smooth muscle cells. *The FASEB Journal*, *13*(1), 103–113. http://www.ncbi.nlm.nih.gov/entrez/query.fcgi?cmd=Retrieve&db=PubMed&dopt=Citation&list_uids=9872935.

Patriarchi, T., Qian, H., Di Biase, V., Malik, Z. A., Chowdhury, D., Price, J. L., et al. (2016). Phosphorylation of Cav1.2 on S1928 uncouples the L-type Ca2+ channel from the beta2 adrenergic receptor. *The EMBO Journal*, *35*(12), 1330–1345. https://doi.org/10.15252/embj.201593409.

Payne, G. W., & Smeda, J. S. (2002). Cerebrovascular alterations in pressure and protein kinase C-mediated constriction in Dahl salt-sensitive rats. *Journal of Hypertension*, *20*(7), 1355–1363. http://www.ncbi.nlm.nih.gov/entrez/query.fcgi?cmd=Retrieve&db=PubMed&dopt=Citation&list_uids=12131532.

Pesic, A., Madden, J. A., Pesic, M., & Rusch, N. J. (2004). High blood pressure upregulates arterial L-type Ca2 + channels. Is membrane depolarization the signal? *Circulation Research*, *94*(10), e97–104. http://www.ncbi.nlm.nih.gov/entrez/query.fcgi?cmd=Retrieve&db=PubMed&dopt=Citation&list_uids=15131006.

Pinho, J. F., Medeiros, M. A., Capettini, L. S., Rezende, B. A., Campos, P. P., Andrade, S. P., et al. (2010). Phosphatidylinositol 3-kinase-delta up-regulates L-type Ca2+ currents and increases vascular contractility in a mouse model of type 1 diabetes. *British Journal of Pharmacology*, *161*(7), 1458–1471. https://doi.org/10.1111/j.1476-5381.2010.00955.x.

Porter, V. A., Bonev, A. D., Knot, H. J., Heppner, T. J., Stevenson, A. S., Kleppisch, T., et al. (1998). Frequency modulation of Ca2+ sparks is involved in regulation of arterial diameter by cyclic nucleotides. *The American Journal of Physiology*, *274*(5 Pt. 1), C1346–C1355. http://www.ncbi.nlm.nih.gov/pubmed/9612222.

Prada, M. P., Syed, A. U., Buonarati, O. R., Reddy, G. R., Nystoriak, M. A., Ghosh, D., et al. (2019). A Gs-coupled purinergic receptor boosts Ca2+ influx and vascular contractility during diabetic hyperglycemia. *eLife*, *8*, e42214. https://doi.org/10.7554/eLife.42214.

Prada, M. P., Syed, A. U., Reddy, G. R., Martin-Aragon Baudel, M., Flores-Tamez, V. A., Sasse, K. C., et al. (2020). AKAP5 complex facilitates purinergic modulation of vascular L-type Ca2+ channel CaV1.2. *Nature Communications*, *11*(1), 5303. https://doi.org/10.1038/s41467-020-18947-y.

Pratt, P. F., Bonnet, S., Ludwig, L. M., Bonnet, P., & Rusch, N. J. (2002). Upregulation of L-type Ca^{2+} channels in mesenteric and skeletal arteries of SHR. *Hypertension, 40*(2), 214–219. http://www.ncbi.nlm.nih.gov/entrez/query.fcgi?cmd=Retrieve&db=PubMed&dopt=Citation&list_uids=12154116.

Qian, H., Patriarchi, T., Price, J. L., Matt, L., Lee, B., Nieves-Cintron, M., et al. (2017). Phosphorylation of Ser1928 mediates the enhanced activity of the L-type Ca2+ channel Cav1.2 by the beta2-adrenergic receptor in neurons. *Science Signaling, 10*(463), eaaf9659. https://doi.org/10.1126/scisignal.aaf9659.

Rangel, E. B., Rodrigues, C. O., & de Sa, J. R. (2019). Micro- and macrovascular complications in diabetes mellitus: Preclinical and clinical studies. *Journal Diabetes Research, 2019*, 2161085. https://doi.org/10.1155/2019/2161085.

Reimer, D., Huber, I. G., Garcia, M. L., Haase, H., & Striessnig, J. (2000). Beta subunit heterogeneity of L-type Ca(2+) channels in smooth muscle tissues. *FEBS Letters, 467*(1), 65–69. http://www.ncbi.nlm.nih.gov/entrez/query.fcgi?cmd=Retrieve&db=PubMed&dopt=Citation&list_uids=10664458.

Roybal, D., Hennessey, J. A., & Marx, S. O. (2020). The quest to identify the mechanism underlying adrenergic regulation of cardiac Ca(2+) channels. *Channels (Austin, Tex.), 14*(1), 123–131. https://doi.org/10.1080/19336950.2020.1740502.

Ruan, X., & Arendshorst, W. J. (1996). Role of protein kinase C in angiotensin II-induced renal vasoconstriction in genetically hypertensive rats. *The American Journal of Physiology, 270*(6 Pt. 2), F945–F952. http://www.ncbi.nlm.nih.gov/entrez/query.fcgi?cmd=Retrieve&db=PubMed&dopt=Citation&list_uids=8764313.

Ruiz-Velasco, V., Zhong, J., Hume, J. R., & Keef, K. D. (1998). Modulation of Ca2+ channels by cyclic nucleotide cross activation of opposing protein kinases in rabbit portal vein. *Circulation Research, 82*(5), 557–565. http://www.ncbi.nlm.nih.gov/entrez/query.fcgi?cmd=Retrieve&db=PubMed&dopt=Citation&list_uids=9529160.

Salamanca, D. A., & Khalil, R. A. (2005). Protein kinase C isoforms as specific targets for modulation of vascular smooth muscle function in hypertension. *Biochemical Pharmacology, 70*(11), 1537–1547. http://www.ncbi.nlm.nih.gov/entrez/query.fcgi?cmd=Retrieve&db=PubMed&dopt=Citation&list_uids=16139252.

Sanderson, J. L., & Dell'Acqua, M. L. (2011). AKAP signaling complexes in regulation of excitatory synaptic plasticity. *The Neuroscientist, 17*(3), 321–336. https://doi.org/10.1177/1073858410384740.

Seki, T., Yokoshiki, H., Sunagawa, M., Nakamura, M., & Sperelakis, N. (1999). Angiotensin II stimulation of Ca^{2+}-channel current in vascular smooth muscle cells is inhibited by lavendustin-A and LY-294002. *Pflügers Archiv, 437*(3), 317–323. http://www.ncbi.nlm.nih.gov/entrez/query.fcgi?cmd=Retrieve&db=PubMed&dopt=Citation&list_uids=9914387.

Shi, Y., & Vanhoutte, P. M. (2017). Macro- and microvascular endothelial dysfunction in diabetes. *Journal of Diabetes, 9*(5), 434–449. https://doi.org/10.1111/1753-0407.12521.

Soldatov, N. M. (1994). Genomic structure of human L-type Ca2+ channel. *Genomics, 22*(1), 77–87. https://doi.org/10.1006/geno.1994.1347.

Somlyo, A. P., & Somlyo, A. V. (2003). Ca^{2+} sensitivity of smooth muscle and nonmuscle myosin II: Modulated by G proteins, kinases, and myosin phosphatase. *Physiological Reviews, 83*(4), 1325–1358. http://www.ncbi.nlm.nih.gov/entrez/query.fcgi?cmd=Retrieve&db=PubMed&dopt=Citation&list_uids=14506307.

Sonkusare, S., Palade, P. T., Marsh, J. D., Telemaque, S., Pesic, A., & Rusch, N. J. (2006). Vascular calcium channels and high blood pressure: Pathophysiology and therapeutic implications. *Vascular Pharmacology, 44*(3), 131–142. doi: S1537-1891(05)00243-0 [pii] https://doi.org/10.1016/j.vph.2005.10.005.

Stehouwer, C. D. A. (2018). Microvascular dysfunction and hyperglycemia: A vicious cycle with widespread consequences. *Diabetes, 67*(9), 1729–1741. https://doi.org/10.2337/dbi17-0044.

Syed, A. U., Reddy, G. R., Ghosh, D., Prada, M. P., Nystoriak, M. A., Morotti, S., et al. (2019). Adenylyl cyclase 5-generated cAMP controls cerebral vascular reactivity during diabetic hyperglycemia. *The Journal of Clinical Investigation*, *129*(8), 3140–3152. https://doi.org/10.1172/JCI124705.

Takahashi, S. X., Miriyala, J., Tay, L. H., Yue, D. T., & Colecraft, H. M. (2005). A CaVbeta SH3/guanylate kinase domain interaction regulates multiple properties of voltage-gated Ca2+ channels [Research Support, N.I.H., Extramural Research Support, Non-U.S. Gov't Research Support, U.S. Gov't, P.H.S.]. *The Journal of General Physiology*, *126*(4), 365–377. https://doi.org/10.1085/jgp.200509354.

Tang, Z. Z., Hong, X., Wang, J., & Soong, T. W. (2007). Signature combinatorial splicing profiles of rat cardiac- and smooth-muscle Cav1.2 channels with distinct biophysical properties. *Cell Calcium*, *41*(5), 417–428. https://doi.org/10.1016/j.ceca.2006.08.002.

Tang, Z. Z., Liang, M. C., Lu, S., Yu, D., Yu, C. Y., Yue, D. T., et al. (2004). Transcript scanning reveals novel and extensive splice variations in human l-type voltage-gated calcium channel, Cav1.2 alpha1 subunit. *The Journal of Biological Chemistry*, *279*(43), 44335–44343. https://doi.org/10.1074/jbc.M407023200.

Tang, Z. Z., Liao, P., Li, G., Jiang, F. L., Yu, D., Hong, X., et al. (2008). Differential splicing patterns of L-type calcium channel Cav1.2 subunit in hearts of spontaneously hypertensive rats and wistar kyoto rats. *Biochimica et Biophysica Acta*, *1783*(1), 118–130. https://doi.org/10.1016/j.bbamcr.2007.11.003.

Tewari, K., & Simard, J. M. (1994). Protein kinase A increases availability of calcium channels in smooth muscle cells from guinea pig basilar artery. *Pflügers Archiv*, *428*(1), 9–16. http://www.ncbi.nlm.nih.gov/pubmed/7526339.

Tiwari, S., Zhang, Y., Heller, J., Abernethy, D. R., & Soldatov, N. M. (2006). Atherosclerosis-related molecular alteration of the human CaV1.2 calcium channel alpha1C subunit. *Proceedings of the National Academy of Sciences of the United States of America*, *103*(45), 17024–17029. https://doi.org/10.1073/pnas.0606539103.

Triggle, D. J. (1991). Calcium-channel drugs: Structure-function relationships and selectivity of action. *Journal of Cardiovascular Pharmacology*, *18*(Suppl. 10), S1–S6. https://www.ncbi.nlm.nih.gov/pubmed/1724996.

Ungvari, Z., Pacher, P., Kecskemeti, V., Papp, G., Szollar, L., & Koller, A. (1999). Increased myogenic tone in skeletal muscle arterioles of diabetic rats. Possible role of increased activity of smooth muscle Ca2+ channels and protein kinase C. *Cardiovascular Research*, *43*(4), 1018–1028. https://doi.org/10.1016/s0008-6363(99)00106-6.

Vaca, L. (2010). SOCIC: The store-operated calcium influx complex. *Cell Calcium*, *47*(3), 199–209. https://doi.org/10.1016/j.ceca.2010.01.002.

Van Petegem, F., Clark, K. A., Chatelain, F. C., & Minor, D. L., Jr. (2004). Structure of a complex between a voltage-gated calcium channel beta-subunit and an alpha-subunit domain. *Nature*, *429*(6992), 671–675. https://doi.org/10.1038/nature02588.

Wamhoff, B. R., Bowles, D. K., & Owens, G. K. (2006). Excitation-transcription coupling in arterial smooth muscle. *Circulation Research*, *98*(7), 868–878. https://doi.org/10.1161/01.RES.0000216596.73005.3c.

Wang, W. Z., Saada, N., Dai, B., Pang, L., & Palade, P. (2006). Vascular-specific increase in exon 1B-encoded CAV1.2 channels in spontaneously hypertensive rats. *American Journal of Hypertension*, *19*(8), 823–831. https://doi.org/10.1016/j.amjhyper.2006.01.020.

Webb, J. G., Yates, P. W., Yang, Q., Mukhin, Y. V., & Lanier, S. M. (2001). Adenylyl cyclase isoforms and signal integration in models of vascular smooth muscle cells. *American Journal of Physiology. Heart and Circulatory Physiology*, *281*(4), H1545–H1552. https://doi.org/10.1152/ajpheart.2001.281.4.H1545.

Weber, S., Bernhard, D., Lukowski, R., Weinmeister, P., Worner, R., Wegener, J. W., et al. (2007). Rescue of cGMP kinase I knockout mice by smooth muscle specific expression of either isozyme. *Circulation Research*, *101*(11), 1096–1103. https://doi.org/10.1161/CIRCRESAHA.107.154351.

Welling, A., Ludwig, A., Zimmer, S., Klugbauer, N., Flockerzi, V., & Hofmann, F. (1997). Alternatively spliced IS6 segments of the alpha 1C gene determine the tissue-specific dihydropyridine sensitivity of cardiac and vascular smooth muscle L-type Ca2+ channels. *Circulation Research*, *81*(4), 526–532. https://doi.org/10.1161/01.res.81.4.526.
Wellman, G. C., Bonev, A. D., Nelson, M. T., & Brayden, J. E. (1996). Gender differences in coronary artery diameter involve estrogen, nitric oxide, and Ca^{2+}-dependent K^+ channels. *Circulation Research*, *79*(5), 1024–1030. http://www.ncbi.nlm.nih.gov/htbin-post/Entrez/query?db=m&form=6&dopt=r&uid=8888695.
Wellman, G. C., Santana, L. F., Bonev, A. D., & Nelson, M. T. (2001). Role of phospholamban in the modulation of arterial Ca^{2+} sparks and Ca^{2+}-activated K^+ channels by cAMP. *American Journal of Physiology. Cell Physiology*, *281*(3), C1029–C1037. http://www.ncbi.nlm.nih.gov/htbin-post/Entrez/query?db=m&form=6&dopt=r&uid=11502581. http://ajpcell.physiology.org/cgi/content/full/281/3/C1029. http://ajpcell.physiology.org/cgi/content/abstract/281/3/C1029.
White, R. E., & Carrier, G. O. (1990). Vascular contraction induced by activation of membrane calcium ion channels is enhanced in streptozotocin-diabetes. *The Journal of Pharmacology and Experimental Therapeutics*, *253*(3), 1057–1062. http://www.ncbi.nlm.nih.gov/entrez/query.fcgi?cmd=Retrieve&db=PubMed&dopt=Citation&list_uids=1694242.
WHO. (2021). *Diabetes*. https://www.who.int/news-room/fact-sheets/detail/diabetes.
Williams, B., Gallacher, B., Patel, H., & Orme, C. (1997). Glucose-induced protein kinase C activation regulates vascular permeability factor mRNA expression and peptide production by human vascular smooth muscle cells in vitro. *Diabetes*, *46*(9), 1497–1503. http://www.ncbi.nlm.nih.gov/entrez/query.fcgi?cmd=Retrieve&db=PubMed&dopt=Citation&list_uids=9287052.
Williams, B., & Schrier, R. W. (1992). Characterization of glucose-induced in situ protein kinase C activity in cultured vascular smooth muscle cells. *Diabetes*, *41*(11), 1464–1472. http://www.ncbi.nlm.nih.gov/entrez/query.fcgi?cmd=Retrieve&db=PubMed&dopt=Citation&list_uids=1397722.
Witczak, C. A., Wamhoff, B. R., & Sturek, M. (2006). Exercise training prevents Ca2+ dysregulation in coronary smooth muscle from diabetic dyslipidemic yucatan swine. *Journal of Applied Physiology (Bethesda, MD: 1985)*, *101*(3), 752–762. https://doi.org/10.1152/japplphysiol.00235.2006.
Xiong, Z., & Sperelakis, N. (1995). Regulation of L-type calcium channels of vascular smooth muscle cells. *Journal of Molecular and Cellular Cardiology*, *27*(1), 75–91. http://www.ncbi.nlm.nih.gov/pubmed/7760390. http://www.jmmc-online.com/article/S0022-2828(08)80009-0/abstract.
Xiong, Z., Sperelakis, N., & Fenoglio-Preiser, C. (1994). Regulation of L-type calcium channels by cyclic nucleotides and phosphorylation in smooth muscle cells from rabbit portal vein. *Journal of Vascular Research*, *31*(5), 271–279. http://www.ncbi.nlm.nih.gov/pubmed/7918917.
Yang, Y., Chen, X., Margulies, K., Jeevanandam, V., Pollack, P., Bailey, B. A., et al. (2000). L-type Ca2+ channel alpha 1c subunit isoform switching in failing human ventricular myocardium. *Journal of Molecular and Cellular Cardiology*, *32*(6), 973–984. https://doi.org/10.1006/jmcc.2000.1138.
Yang, L., Katchman, A., Morrow, J. P., Doshi, D., & Marx, S. O. (2011). Cardiac L-type calcium channel (Cav1.2) associates with gamma subunits. *The FASEB Journal*, *25*(3), 928–936. https://doi.org/10.1096/fj.10-172353.
Yeow, S. Q. Z., Loh, K. W. Z., & Soong, T. W. (2021). *Calcium channel splice variants and their effects in brain and cardiovascular function* (pp. 67–86). Singapore: Springer. https://doi.org/10.1007/978-981-16-4254-8_5.

CHAPTER FIVE

Multiphasic changes in smooth muscle Ca^{2+} transporters during the progression of coronary atherosclerosis

Jill Badin[a], Stacey Rodenbeck[b], Mikaela L. McKenney-Drake[c], and Michael Sturek[d,*]

[a]ZOLL Medical Corporation, Chelmsford, MA, United States
[b]Department of Biology, Harding University, Searcy, AR, United States
[c]Butler University, Health Sciences Department, Pharmacy and Health Sciences, Indianapolis, IN, United States
[d]Department of Anatomy, Cell Biology, & Physiology, Indiana University School of Medicine, Indianapolis, IN, United States
*Corresponding author: e-mail address: msturek@iu.edu

Contents

Abstract

Ischemic heart disease due to macrovascular atherosclerosis and microvascular dysfunction is the major cause of death worldwide and the unabated increase in metabolic syndrome is a major reason why this will continue. Intracellular free Ca^{2+} ($[Ca^{2+}]_i$) regulates a variety of cellular functions including contraction, proliferation, migration, and

Current Topics in Membranes, Volume 90
ISSN 1063-5823
https://doi.org/10.1016/bs.ctm.2022.09.007

transcription. It follows that studies of vascular Ca^{2+} regulation in reductionist models and translational animal models are vital to understanding vascular health and disease. Swine with metabolic syndrome (MetS) develop the full range of coronary atherosclerosis from mild to severe disease. Intravascular imaging enables quantitative measurement of atherosclerosis in vivo, so viable coronary smooth muscle (CSM) cells can be dispersed from the arteries to enable Ca^{2+} transport studies in native cells. Transition of CSM from the contractile phenotype in the healthy swine to the proliferative phenotype in mild atherosclerosis was associated with increases in SERCA activity, sarcoplasmic reticulum Ca^{2+}, and voltage-gated Ca^{2+} channel function. In vitro organ culture confirmed that SERCA activation induces CSM proliferation. Transition from the proliferative to a more osteogenic phenotype was associated with decreases in all three Ca^{2+} transporters. Overall, there was a biphasic change in Ca^{2+} transporters over the progression of atherosclerosis in the swine model and this was confirmed in CSM from failing explanted hearts of humans. A major determinant of endolysosome content in human CSM is the severity of atherosclerosis. In swine CSM endolysosome Ca^{2+} release occurred through the TPC2 channel. We propose a multiphasic change in Ca^{2+} transporters over the progression of coronary atherosclerosis.

1. Overview of vascular smooth muscle intracellular Ca^{2+} regulation in health

Ca^{2+} is arguably one of the most important signaling molecules in coronary smooth muscle (CSM) (Amberg & Navedo, 2013) and the concentration of intracellular free Ca^{2+} ($[Ca^{2+}]_i$) regulates a variety of cellular functions including contraction, proliferation, migration, and transcription (House, Potier, Bisaillon, Singer, & Trebak, 2008; Kudryavtseva, Aalkjaer, & Matchkov, 2013; Lundberg, Curto, Bilato, Monticone, & Crow, 1998; Owens, Kumar, & Wamhoff, 2004; Pauly et al., 1995; Wamhoff et al., 2004).

Fig. 1 shows major Ca^{2+} transporters that are the main focus of this chapter. Influx of Ca^{2+} into CSM from the extracellular space occurs largely via voltage-gated Ca^{2+} channels (VGCCs) (Jackson, 2020; Moosmang et al., 2003; Ottolini & Sonkusare, 2021; Sturek, 2011). Ligand-gated transient receptor potential (TRP) channels (Dineen, Neeb, Obukhov, & Sturek, 2016; Jackson, 2020; Nishida, Kuwahara, Kozai, Sakaguchi, & Mori, 2015; Sturek, Mokelke, Sindermann, Adam, & March, 2006) and Orai1 channels (Climent et al., 2020; Ng et al., 2012; Zhang et al., 2011) are activated by ligand-binding to an extracellular receptor coupled to the channel or by other second messengers or depletion of the sarcoplasmic reticulum (SR) Ca^{2+} store that results in so-called store-operated Ca^{2+} entry (SOCE). The Ca^{2+}-sensing proteins on the SR membrane called stromal interaction

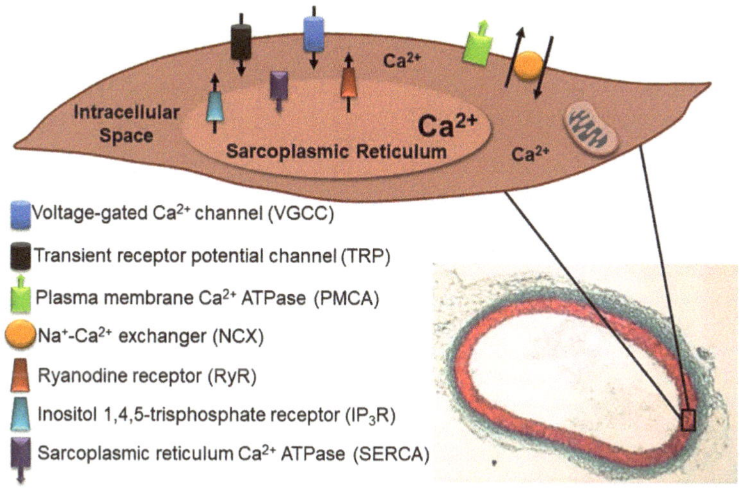

Fig. 1 Overview of Ca^{2+} transporters in vascular smooth muscle of the healthy artery. Vascular smooth muscle cells make up the media (red layer lower right image) in this healthy coronary artery. Connective tissue is shown in the blue stain of the perimeter of the artery. Major Ca^{2+} transporters are shown with abbreviations. Subtypes and different isoforms are not shown for simplicity. The sarcoplasmic reticulum (SR) is the major intracellular Ca^{2+} store. The rightmost structure in the model cell is a mitochondrion.

molecule 1 (STIM1), upon detection of SR Ca^{2+} store depletion, migrate to SR-plasma membrane junctions to form clusters and recruit Orai1 channels on the plasma membrane (Ong, de Souza, & Ambudkar, 2016; Zhang et al., 2005). This STIM1-Orai1 complex also serves as a Ca^{2+} entry mechanism that contributes to refilling of the SR Ca^{2+} stores (Leung, Yung, Yao, Laher, & Huang, 2008). These ligand-gated channels have a significant role in Ca^{2+} influx, albeit to a lesser extent. Although channel gating for ligand-gated and SOCE is voltage-independent, the ion flux through the channel is dependent on transmembrane driving force, i.e., there is greater Ca^{2+} and Na^{+} entry at polarized membrane potentials. This is clearly different from Ca^{2+} entry through VGCC that requires membrane depolarization. In addition to regulating vascular tone (Leung et al., 2008), SOCE are involved in VSM proliferation (Trebak et al., 2013). As such, SOCE is involved in pathological remodeling in many vascular disorders (Leung et al., 2008; Trebak et al., 2013). The TRPC1 channel that is involved in SOCE in CSM has been linked to proliferation and hypertrophy (Landsberg & Yuan, 2004; Takahashi et al., 2007), neointimal hyperplasia after vascular injury (Kumar et al., 2006), and atherosclerosis (Edwards et al., 2010; Li et al., 2017).

After Ca^{2+} enters the cell and is available in the intracellular space, $[Ca^{2+}]_i$ can be buffered by the sarcoplasmic reticulum (SR) via the sarco-endoplasmic reticulum Ca^{2+}-ATPase (SERCA) (Hill, Dixon, & Sturek, 2001; Hill, Price, Dixon, & Sturek, 2003; Neeb et al., 2010; Sturek, Kunda, & Hu, 1992; Underwood, Laughlin, & Sturek, 1994; Witczak & Sturek, 2004, 2005). The SERCA pump is the only transporter that enables Ca^{2+} uptake by the SR and has been shown to sequester more than 70% of $[Ca^{2+}]_i$ (House et al., 2008; MacLennan, Asahi, & Tupling, 2003). The sequestered Ca^{2+} within the SR can be released in the intracellular space by inositol trisphosphate receptors (IP3R) and ryanodine receptors (RyR) to rapidly increase $[Ca^{2+}]_i$. Efflux of Ca^{2+} occurs via the plasma membrane Ca^{2+} ATPase (PMCA) and Na^+-Ca^{2+} exchanger (NCX) (Fig. 1; (Sturek, 2011)). Due to the importance of $[Ca^{2+}]_i$ in cellular functioning and the complexity and variety of Ca^{2+} channels and transporters present in the plasmalemmal and SR membranes, CSM must integrate multiple signaling events that influence $[Ca^{2+}]_i$ either directly or indirectly. Additionally, the fluxes in $[Ca^{2+}]_i$ can affect how these Ca^{2+} transporters interact with both Ca^{2+} and other elements in the cellular environment.

It is nearly common knowledge that increased CSM $[Ca^{2+}]_i$ to trigger contraction is accomplished primarily by depolarization of the plasma membrane to activate Ca^{2+} influx through voltage-gated Ca^{2+} channels (Klabunde, 2005; Sturek et al., 2006). The local increase in $[Ca^{2+}]_i$ from Ca^{2+} influx through voltage-gated Ca^{2+} channels also can cause the activation of RyR, subsequent release of Ca^{2+} from the SR store, and further increases in $[Ca^{2+}]_i$ (Fernandez-Tenorio et al., 2010). This Ca^{2+}-induced Ca^{2+}-release functions as an amplifying mechanism for Ca^{2+}-mediated signaling in vascular smooth muscle (Fernandez-Tenorio et al., 2010), but has a negligible role in excitation-contraction coupling compared to that in cardiac muscle. In vascular smooth muscle the typically lower VGCC density, lesser inactivation of VGCC, and sustained depolarization fuel sustained contractions (Fernandez-Tenorio et al., 2010; Sturek et al., 2006). In contrast, robust and transient SR Ca^{2+} release elicited by pharmacological agents (e.g., caffeine) typically elicits only transient contractions.

2. Smooth muscle phenotypic modulation and $[Ca^{2+}]_i$ handling alterations

Differentiated, contractile CSM populate the tunica media of coronary arteries shown in Fig. 1 and typically have a low proliferation rate (Owens et al., 2004). However, CSM can dedifferentiate into a proliferative, migratory phenotype that secretes extracellular matrix (ECM) proteins as a wound-healing response (Owens et al., 2004). Because coronary artery disease (CAD) is associated with constant inflammation, oxidative stress, and cellular stress, CSM continually dedifferentiate into other phenotypes, including osteogenic phenotypes and the previously mentioned migratory/synthetic phenotypes contributing to plaque development and growth, arterial calcification, and overall disease pathology (Owens et al., 2004; Sturek, 2011).

Cellular responses to changes in $[Ca^{2+}]_i$ often underlie our definition of what a given cell "is" (Hille, 1992; House et al., 2008). As $[Ca^{2+}]_i$ controls a litany of cellular functions, phenotypic modulation is closely coupled to changes in $[Ca^{2+}]_i$ handling in a process called "excitation-transcription coupling." Many transcription factors such as serum response factor (SRF), cAMP response element binding (CREB), and nuclear factor of activated T-cells (NFAT) are Ca^{2+}-dependent (House et al., 2008; Sturek, 2011). Ca^{2+} signaling in CSM depends on local $[Ca^{2+}]_i$ signaling events such as Ca^{2+} sparks, waves, and wavelets with low global $[Ca^{2+}]_i$ (Amberg & Navedo, 2013; Sturek, 2011). The SERCA pump and subsequent SR Ca^{2+} sequestration facilitates this low global $[Ca^{2+}]_i$ level. Substantial local increases in $[Ca^{2+}]_i$ activate SRF, which binds to the promoter region of various genes involved in contractile phenotype maintenance (Owens et al., 2004). A dysfunctional SERCA prevents the SR from being able to sequester Ca^{2+}, thereby leading to greater SOCE and in some cases a sustained increase in global $[Ca^{2+}]_i$ (Berra-Romani, Mazzocco-Spezzia, Pulina, & Golovina, 2008; House et al., 2008; Matchkov, Kudryavtseva, & Aalkjaer, 2012). High global $[Ca^{2+}]_i$ activates Ca^{2+}/calmodulin-dependent protein kinase II, which in turn phosphorylates CREB. Once phosphorylated CREB is translocated into the nucleus, it serves as a transcription factor for genes involved in proliferation, migration, and survival (Berra-Romani et al., 2008). High global $[Ca^{2+}]_i$ also increases the activation of calcineurin, which activates NFAT through dephosphorylation and promotes the transcription of genes involved in inflammation and immune response (Matchkov et al., 2012). In this way, $[Ca^{2+}]_i$ handling patterns regulate gene expression and, ultimately, cell phenotype and identity.

3. Metabolic syndrome "milieu"

Metabolic syndrome (MetS) is commonly defined as the clustering together of at least three of the following risk factors: central obesity, glucose intolerance, insulin resistance, hypertension, hypertriglyceridemia, and hypercholesterolemia (Roberts, Hevener, & Barnard, 2013). MetS has long been known to associate with cardiovascular disease, but until recently, its causal role in the early development of cardiovascular disease remained unknown. In 2002, a longitudinal, prospective cohort study of 1209 Finnish men between the ages of 42 and 60 years at baseline revealed that the presence of MetS prior to incidence of cardiovascular disease increased the risk of mortality due to cardiovascular disease in general and due to coronary heart disease specifically (Lakka et al., 2002). More recently, Kim et al. sought to determine the contributions of individual MetS risk factors in the development of CAD (Kim et al., 2010). While this non-randomized, retrospective study found no change in the prevalence of CAD with regards to the presence of MetS, CAD severity was increased by the greater clustering of MetS-associated risk factors. Most recently, a meta-analysis of 10 observational studies indicated that presence of MetS or any of its risk factors is significantly associated with a high risk for CAD (Alshammary, Alharbi, Alshehri, Vennu, & Ali Khan, 2021). It is likely that the admittedly heterogeneous MetS milieu may synergistically affect CAD, making it difficult to isolate the contributions of individual risk factors. However, studies have examined individual parameters of MetS either directly or indirectly, which may shed light on the individual contributions. We focus here on the potential effects of the constellation of components in vivo in MetS on $[Ca^{2+}]_i$ regulation in CSM. The hallmarks of MetS are induced in Ossabaw miniature swine using an experimental design timeline as shown in Fig. 2 to produce varying severities of atherosclerosis from which we have characterized Ca^{2+} transporter dysfunction. Pigs are fed an atherogenic diet that is high in calories, fructose, fat, and cholesterol for 6–12 months, which elicits MetS as shown in numerous studies cited in our recent review (Sturek, Alloosh, & Sellke, 2020). The pigs progress from a "normal, healthy" lean state at birth. The body weight gain (left y-axis) is relatively representative of the growth curves for Ossabaw swine. The age of the pigs is critical for the developmental stage, as we are studying adult, sexually mature pigs, not juvenile, pediatric subjects in the 0–6 month age range. This "normal," healthy status is also depicted as near minimum (min) relative insulin resistance, etc., on the right axis.

Hyperglycemia, acting synergistically with dyslipidemia, may well sit at the head of the cascade of events leading from MetS to CAD. Hyperglycemia is a hallmark of type 2 diabetes and some definitions of MetS require its presence. Hyperglycemia is implicated in the hypercontractile state of resistance arteries and arterioles observed in vascular dysfunction. The hyperglycemic state has

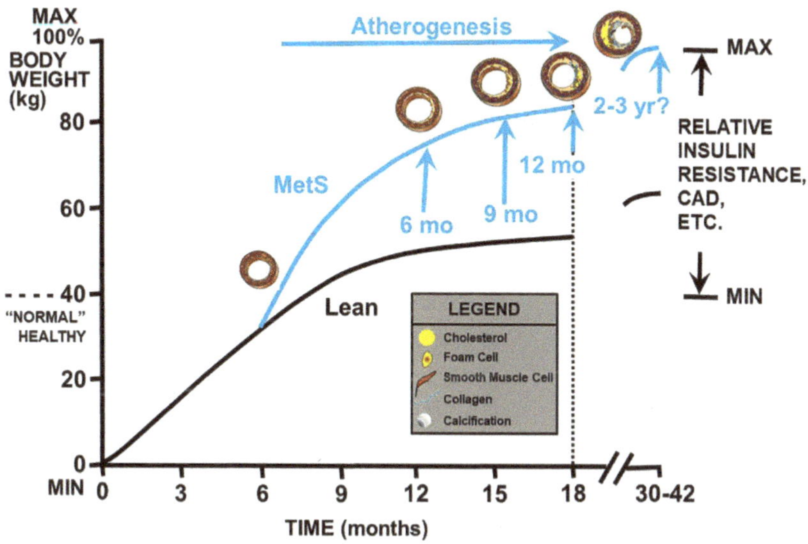

Fig. 2 Experimental design timeline for eliciting metabolic syndrome and studying Ca^{2+} transporters during the progression of atherosclerosis. Ossabaw miniature swine have been systematically studied after 6, 9, and 12 months on an atherogenic diet. These three groups of swine are compared to a healthy, lean group. The illustrations show typical atherosclerosis at the time points. The 2–3 year duration of atherogenesis is just a hypothetical study, not that which is used for data to be discussed. Characteristics of metabolic syndrome, e.g., insulin resistance, coronary artery disease (CAD), are shown to go from the minimum (MIN) in the "normal, healthy," lean pig to some maximum (MAX) value. See text for more details.

long been known to cause endothelial dysfunction, but several more recent studies have undertaken the question of whether it has more direct effects on the underlying smooth muscle. Because the observed changes in vascular smooth muscle function in diabetes rely on protein signaling products that share alterations in $[Ca^{2+}]_i$ as an upstream regulator, Searls, Loganathan, Smirnova, and Stehno-Bittel (2010) examined the effect of diabetic conditions on a number of intracellular $[Ca^{2+}]_i$ transporters. In SMCs from the aortas and femoral arteries of two diabetic mouse lines, IP_3-mediated release of Ca^{2+} from intracellular stores was blunted. These findings were recapitulated in an acute examination of A7r5 SMCs cultured in high glucose media in an effort to mimic hyperglycemic conditions. Further examination revealed that overall IP_3R and SERCA expression was reduced and the remaining IP_3R and SERCA were redistributed to the nuclear membrane. Searls et al. did not address mechanisms leading to the reduction and redistribution of SERCA and IP_3R and did not examine effects on other Ca^{2+} transport mechanisms.

4. Atherogenesis

The MetS and diabetic "milieu" elicit a cascade of events in the vascular wall that result in atherosclerosis as reviewed recently by Libby and colleagues (Libby, 2021). The arterial wall during atherogenesis is illustrated in several simplified schematics of arterial cross-sections in Fig. 2 and histology of cross-sections of coronary arteries from both humans and swine in various stages of disease severity is shown later in this chapter. Early stages of atherosclerosis include fatty streaks and preatheroma development that is characterized by outward/positive remodeling, foam cell infiltration, extracellular lipid deposition, and cell proliferation and migration. Longer duration elicits true atheroma in which a lipid core is formed and there is the start of lumen occlusion. A later stage is the fibroatheroma that has a fibrous cap, lipid core, and a positive coronary artery calcium score as measured by computed tomography scan and intravascular ultrasound. The most developed stage is the complicated lesion that is composed of hard collagen, calcified plaque, and cell debris (McKenney-Drake et al., 2018).

Quantification of atherosclerosis in vivo using intravascular ultrasound is shown in Fig. 3. All three major epicardial arteries (Fig. 3, right; RC, right coronary; LAD, left anterior descending; CFX, circumflex) are interrogated by intravascular ultrasound along the entire artery length to quantify areas for regions with diffuse atherosclerosis and focal stenotic regions (Sturek, 2011). Soft atheroma is visualized as less echogenic than the adventitia. Calcification is detected by a highly echogenic arc that induces "acoustic shadowing" which is drop-out of the intravascular ultrasound signal peripheral to the dense calcification (Fig. 3, left). Plaque burden (PB, % stenosis) is shown by thickening and wall coverage (double arrow arc), which can range from <10% to 100% of the artery inner wall circumference (McKenney-Drake, Rodenbeck, et al., 2016; McKenney-Drake, Territo, et al., 2016; Neeb et al., 2010; Sturek, Tune, & Alloosh, 2015). Our recent study (Badin et al., 2018) quantified "spotty" calcification by image analysis of whole arterial length, mainly detected as "spots" or patches of echogenicity like the small acoustic shadowing shown by the dashed lines and S on the left side of Fig. 3 (Phillips-Eakley et al., 2015). This is consistent with early calcification that was found in that 6 month duration study. Similar methods are used in human studies and spotty coronary artery calcification was found to be unstable, increasing the risk of acute myocardial infarctions (Ehara et al., 2004; Irkle et al., 2015). Major reasons for using this method are: (1) it is

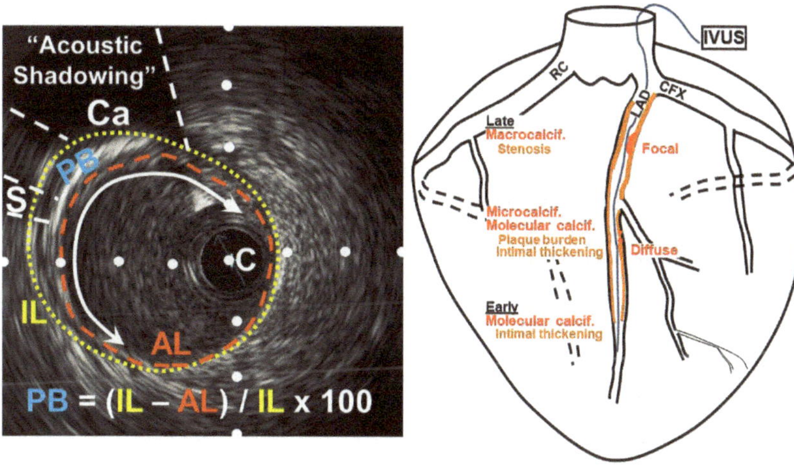

Fig. 3 Intravascular ultrasound measurement of atherosclerosis in vivo. (Left Panel) Plaque burden and calcification are determined by the equation shown, where: IL = initial lumen, AL = actual lumen, S = spotty calcification, PB = plaque burden; all shown in the color-coded perimeters of areas. C = catheter, Ca = large, calcified region. White dots represent 1 mm distance. (Right Panel) A catheter is inserted in all three major epicardial arteries (RC, right coronary; LAD, left anterior descending; CFX, circumflex) via the ascending aorta, and all coronary arteries are interrogated along the entire artery length to quantify atherosclerotic areas. Both early and late coronary disease are discernable this way, as it enables the visualization of both focal and diffuse plaques throughout the length of all coronary arteries.

exceptionally translational, as all devices are used in human patients; and (2) regions of soft atheroma, hard calcification, and healthy arteries can be used selectively for acute dispersion of viable smooth muscle cells for characterization of ion transporters (Sturek, 2011). The cells harvested represent overall the native state of the smooth muscle cell in vivo.

Based on previous sections highlighting the important role of Ca^{2+} in regulating CSM transcription, proliferation and migration, we and others sought to characterize changes in $[Ca^{2+}]i$ and the Ca^{2+} transporters involved. This could drive the search for drug targets. For example, the classic study of Henry and Bentley showed that the voltage-gated Ca^{2+} channel inhibitor nifedipine strongly attenuated atherosclerosis in a cholesterol-fed rabbit model (Henry & Bentley, 1981). Cellular Ca^{2+} overload as a pathogenic mechanism of cell injury became dogma. Our initial hypothesis was that Ca^{2+} transporter dysfunction that leads to $[Ca^{2+}]i$ overload would be directly proportional to the severity of atherosclerosis. Further, the time integral of the increase in $[Ca^{2+}]i$ should be another factor in longer disease duration. Fig. 2 in "Introduction to ion transport and membrane interactions in vascular health and disease" by Sturek shows the temporal relationship for contraction and atherogenesis.

5. Function/dysfunction of Ca^{2+} transporters during coronary atherosclerosis progression in swine

We use fura-2 imaging extensively to study $[Ca^{2+}]i$ regulation. Fig. 4 provides a representative tracing to demonstrate the typical experimental protocol for Ca^{2+} signaling in CSM freshly isolated from a coronary artery or in CSM after organ culture of arteries. Depolarization with 80 mM K+ elicited Ca^{2+} influx through VGCC. A maximal 5 mM caffeine (Caf) dose opened ryanodine receptors and the absence of extracellular Ca^{2+} (including 10 μM EGTA in the superfusate) ensured that the peak response is due to release of Ca^{2+} from the SR Ca^{2+} store. During caffeine exposure in Ca^{2+}-free solution the decay of the $[Ca^{2+}]i$ toward baseline is largely due to Ca^{2+} efflux from the cell. After rapid wash-out of the caffeine upon return to a Ca^{2+}-free solution, the $[Ca^{2+}]i$ decay is due to SERCA functioning to sequester $[Ca^{2+}]i$ to replenish the SR Ca^{2+} store. A striking feature of $[Ca^{2+}]$ i regulation in smooth muscle cells from healthy conduit coronary arteries is

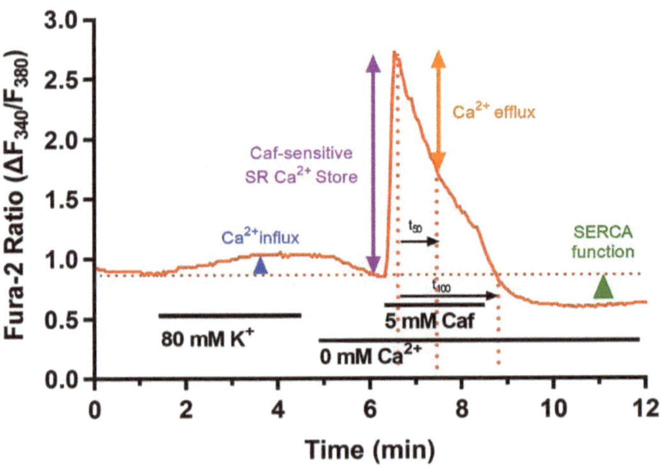

Fig. 4 Typical experimental protocol for Ca^{2+} signaling in CSM freshly dispersed from arteries or organ cultured arteries. Cells were constantly superfused with buffer solutions at a flow rate of 1–2 mL/min. Physiological salt solution was the baseline and other solutions were superfused for the period indicated by the horizontal lines. These solutions manipulated the Ca^{2+} transporters labeled, e.g., Ca^{2+} influx. Abbreviations: Caf, caffeine; t_{50}, time for 50% recovery to baseline; t_{100}, time for 100% recovery to baseline.

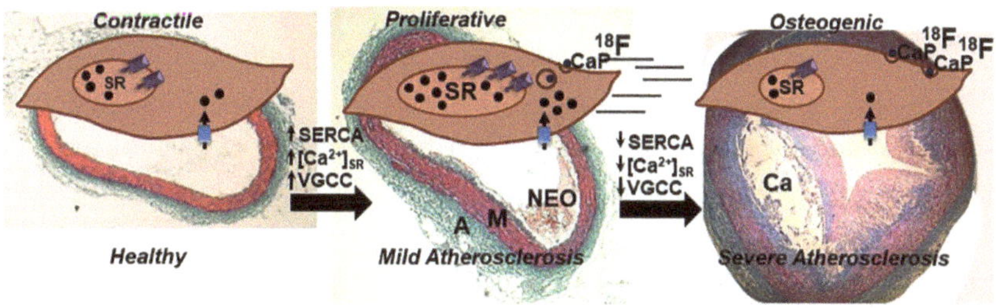

Fig. 5 Intracellular Ca^{2+} regulation is altered in a biphasic manner in proportion to atherosclerosis severity (summarized from McKenney-Drake, Rodenbeck, et al., 2016; McKenney-Drake, Territo, et al., 2016). Transition of coronary smooth muscle cells from the contractile phenotype in the healthy pig (left panel) to the proliferative phenotype in mild atherosclerosis (center panel) was associated with increases in SERCA activity, SR Ca^{2+} store capacity, and voltage-gated Ca^{2+} channel function. Transition from the proliferative to a more osteogenic phenotype (right panel) was associated with decreases in all three parameters. Abbreviations and symbols: SR, sarcoplasmic reticulum; ↑, increase; ↓, decrease; •, Ca^{2+}; SERCA, sarco-endoplasmic reticulum Ca^{2+} ATPase; VGCC, voltage-gated Ca^{2+} channels; Ca, large calcified region; CaP, hydroxyapatite; ^{18}F, radioactive F; A, adventitia; M, media; NEO, neointima; SERCA and VGCC symbols as in Fig. 1.

that the peak increase in $[Ca^{2+}]i$ due to VGCC activation is only about half the peak increase due to Ca^{2+} release from the SR.

Function/dysfunction of Ca^{2+} transporters is summarized schematically in Fig. 5. Healthy coronary smooth muscle cells from lean Ossabaw miniature swine typically present in the contractile phenotype. Transition of coronary smooth muscle cells from the contractile phenotype in the healthy pig to the proliferative phenotype in mild atherosclerosis ("early stage") was associated with increases in SERCA activity, SR Ca^{2+} storage capacity, and voltage-gated Ca^{2+} channel function. Transition from the proliferative to a more osteogenic phenotype ("late stage") was associated with decreases in SERCA activity, SR Ca^{2+} storage capacity and voltage-gated Ca^{2+} channel function. Note the histology staining of porcine coronary arteries in different stages of CAD severity presented in Fig. 5. Our recent data indicate that early calcification exists in the type 2 fatty streak and preatheroma (McKenney-Drake et al., 2018; McKenney-Drake, Rodenbeck, et al., 2016; McKenney-Drake, Territo, et al., 2016), which have only a thin region of neointimal thickening (NEO) separated by the thin internal elastic lamina. The lumen is enlarged (outward positive remodeling). The medial (M) layer shows some thickening due to proliferation of the coronary smooth muscle and the adventitia (A) is clearly demarcated from the other structures of the artery by the blue connective tissue in the Trichrome image.

Support of outward positive remodeling in early coronary artery disease was provided by computed tomography coronary angiography in which a lean Ossabaw swine is compared to a MetS Ossabaw after only 4 months of atherogenic diet-induced MetS (Choy et al., 2015). The example of the "osteogenic" phenotype in Fig. 5 is extreme with a necrotic core. The necrotic core shown in late atherosclerosis in Fig. 5 was corroborated by the near abolishment of Ki-67 staining that indicates little to no active proliferation (McKenney-Drake, Rodenbeck, et al., 2016; McKenney-Drake, Territo, et al., 2016), which is consistent with apoptosis.

6. SERCA stimulation induces coronary smooth muscle proliferation in early atherosclerosis

The increase in medial area in early atherosclerosis (Fig. 5) is consistent with the role of increased SERCA activity in proliferation, but it does not provide compelling evidence for causation. Our lab used the selective SERCA activator CDN1163 (Kang et al., 2016) along with the SERCA inhibitor, cyclopiazonic acid (CPA) to test for causation in vitro (Fig. 6). This measure of acute activation of SERCA lead to effects of chronic activation on smooth muscle growth and cell proliferation in our organ culture model (Edwards et al., 2008; Shen, Halenda, Sturek, & Wilden, 2005; Wamhoff, Bowles, Dietz, Hu, & Sturek, 2002).

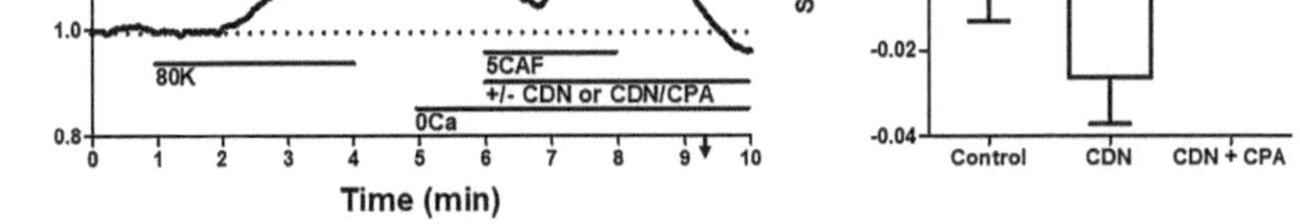

Fig. 6 SERCA activity can be increased and decreased pharmacologically in coronary smooth muscle. Either 10 μM CDN1163 or 10 μM CDN1163 (CDN) plus 10 μM cyclopiazonic acid (CPA) were added to the superfusate to determine SERCA activity. *CDN < CDN + CPA, $p < 0.05$.

6.1 Methods

6.1.1 Organ culture of epicardial conduit coronary arteries

Epicardial coronary arteries were excised under sterile conditions from domestic swine hearts similar to previous reports (Edwards et al., 2008; Shen et al., 2005; Wamhoff et al., 2002). Arteries were cleared of adherent connective tissue and sectioned into arterial rings of 2–4 mm in length. Rings were cultured for 7 days in 12-well plates at 37 °C with 4.6% CO_2 in a base media of Dulbecco's modified eagle medium (DMEM; Sigma-Aldrich, St. Louis, MO) containing high glucose (180 mg/dL), 1% penicillin streptomycin (PS; Gemini Bio-products, Sacramento, CA) and 20% fetal bovine serum (Gibco, Grand Island, NY). CPA (10 μM) was delivered to the culture medium using dimethyl sulfoxide (DMSO) as a vehicle. Another group was cultured in base media with DMSO as a negative control (Fig. 7, top left). Culture media was changed every 2 days. Cultured arterial rings were compared to freshly isolated coronary arteries as a control.

6.1.2 Proliferation assay

Epicardial conduit coronary arteries were prepared as described above and cultured in 12-well plates at 37 °C with 4.6 CO_2 for 3 days in DMEM

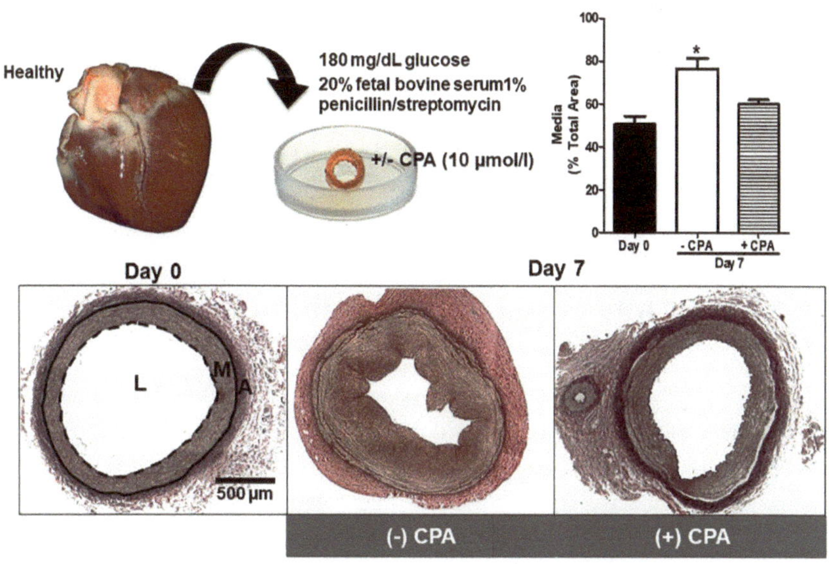

Fig. 7 SERCA inhibition blocks medial thickening in an organ culture model of coronary artery disease. Abbreviations: CPA, cyclopiazonic acid. −CPA > Day 0 and +CPA, $p < 0.05$.

containing normal glucose (100 mg/dL) and 1% PS in the absence of FBS (serum starved conditions). Then, fresh media was added and the culture was maintained for 1 day in the presence of the selective SERCA activator CDN1163 (10 μM) and in the presence or absence of CPA (10 μM). Control rings were cultured for the final day in serum free conditions without CDN1163 or CPA.

6.1.3 Immunohistochemistry

Coronary artery segments embedded in paraffin were transported the Department of Pathology at Indiana University School of Medicine and processed. Immunostaining was performed using Ki-67 as a proliferation marker. Images were captured using a LEICA DM 300 inverted microscope and analyzed with Adobe Photoshop CS6. Positive Ki-67 nuclei are expressed as a percentage of total nuclei.

6.1.4 Statistical analysis

Statistical analysis was performed using GraphPad Prism 5.0 (San Diego, CA). One-way analysis of variance (ANOVA) with student Newman-Keuls post hoc analysis was performed. Data are represented as mean ± SEM. $p < 0.05$ was considered statistically significant.

6.2 Results

Fig. 7 shows that CPA inhibition of SERCA significantly decreased arterial medial area. Stimulation with CDN1163 significantly increased proliferation, as shown by increased Ki-67 expression in Fig. 8. Collectively, these data show a causal role of SERCA activation in proliferation of coronary smooth muscle cells.

7. Similarity of swine and human coronary atherosclerosis and smooth muscle Ca^{2+} signaling

We have observed late coronary artery calcification resolvable with intravascular ultrasound, computed tomography, and histology in MetS (pre-diabetic) Ossabaw miniature swine, while ~120 grossly diabetic dyslipidemic Yucatans showed no coronary artery calcification (Mokelke, Dietz, Eckman, Nelson, & Sturek, 2005; Neeb et al., 2010; Sturek, 2011). Coronary histology of these stages is shown in Fig. 9 (and correspond reasonably well to the mild and severe atherosclerosis stages in Fig. 5). Fig. 9 shows in human coronary artery the progression through several stage of

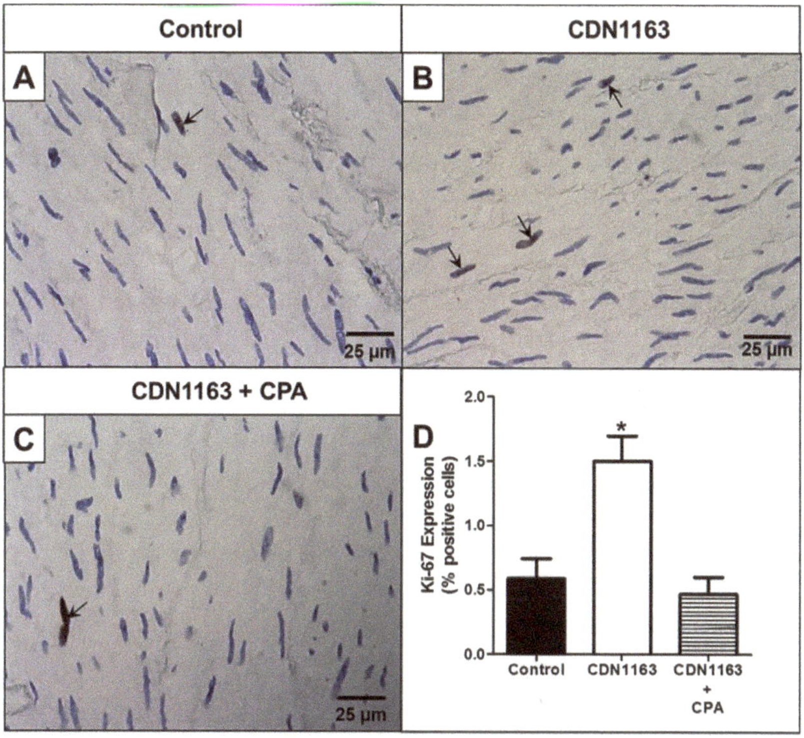

Fig. 8 SERCA activation induces coronary smooth muscle proliferation in organ culture. Arrows show Ki-67 staining. CDN1163 > Control and CDN1163 + CPA, $p < 0.05$.

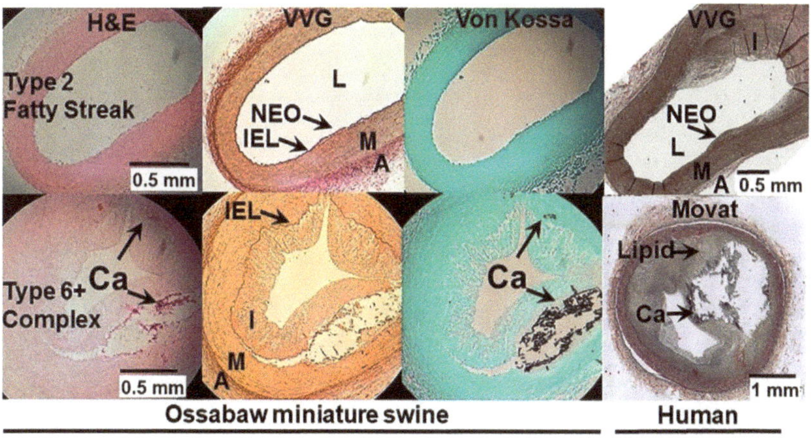

Fig. 9 Similarity of swine and human coronary atherosclerosis. See text for details. Abbreviations: H&E, hematoxylin & eosin; VVG, Verhoeff-Van Giesen; L, lumen; NEO, neointima; IEL, internal elastic lamina; M, media; A, adventitia; I, intima.

atherosclerosis. In the upper right Verhoeff-Van Giesen (VVG) image there is normal and diseased sections. The normal arterial wall is in the lower left corner. The artery transitions to neointimal thickening (NEO) and marked thickened intima (I), more mature plaque burden, and thick medial (M) layer. The lower right image shows in human coronary artery a complex type 6+ atherosclerotic lesion with calcification within a lipid pool (Movat stain). Ossabaw pig arteries show similar lesions on H&E and Von Kossa staining (arrows to dark mass), which are sites of structural, late, macrocalcification. Note the very mild microcalcification site in the upper part of the artery in the H&E and Von Kossa images of the type 6+ complex lesions. These microcalcifications can be seen occasionally in earlier stages of coronary artery disease, but their presence is not statistically significant on histology when rigorously compared to coronary arteries from healthy, lean control pigs (McKenney-Drake, Rodenbeck, et al., 2016; McKenney-Drake, Territo, et al., 2016). The major point is that these swine and human coronary arteries have similar features, which is another reason that studying ion transport in swine coronary arteries may generate data that has high fidelity translation to human medicine.

Figs. 10 and 11 show that the similarities between swine and human coronary arteries go beyond structure to pathophysiological function.

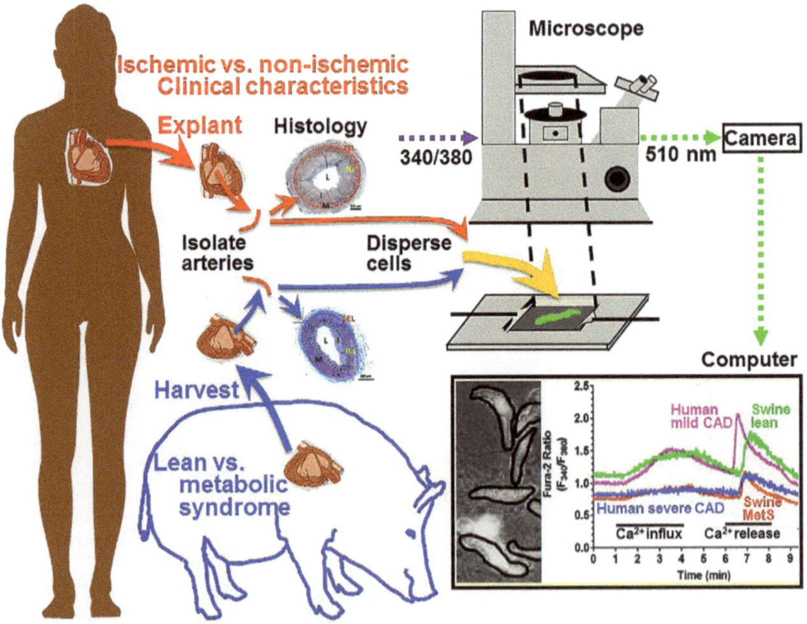

Fig. 10 Similarity of swine and human coronary smooth muscle Ca^{2+} signaling (Badin et al., 2021). See text for details. *Reproduced with permission.*

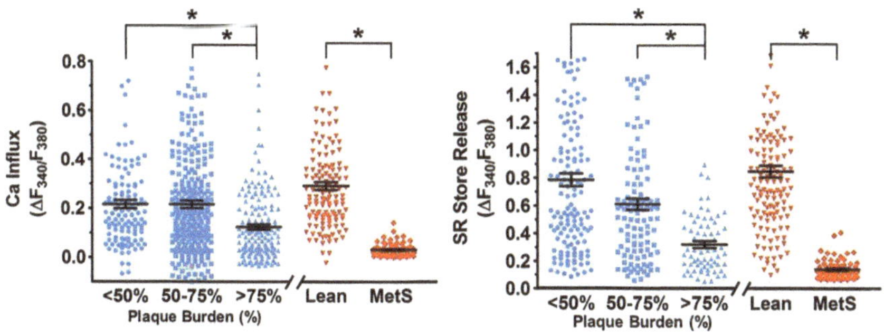

Fig. 11 Scatter plot analysis of voltage-gated Ca^{2+} influx and sarcoplasmic reticulum Ca^{2+} release in swine and humans (Badin et al., 2021). See text for details. *Reproduced with permission.*

We conducted a study over a period of 3 years during which we obtained hearts explanted from 24 patients who were receiving a transplant for ischemic or non-ischemic heart failure. The workflow and major results are shown in Fig. 10. Coronary arteries were isolated from explanted human hearts from patients undergoing heart transplantation surgery and from swine that had been on atherogenic diet for 11 months, smooth muscle cells were dispersed within 24 h of harvest, then Ca^{2+} signaling was studied with our protocol described previously (Fig. 4). The graphic in Fig. 10 shows the phenomenal similarity between humans and swine with mild and severe CAD. More quantitative analysis showing data points for all cells studied from coronary arteries of 24 humans and 14 swine in Fig. 11 shows excellent concordance, but intriguing variability of the data that points to the heterogeneity of patient characteristics. While this is an important result to argue for translation from swine to humans, it is obvious that not every ion transporter and cell signaling pathway will be the same and all should be studied rigorously before drawing broad conclusions.

8. Multiphasic model of coronary smooth muscle Ca^{2+} transporter regulation in atherosclerosis

We originally described the Ca^{2+} transporter changes in the progression of atherosclerosis as biphasic, because the function of VGCC, SR Ca^{2+} store, and SERCA increased in early and decreased in late atherosclerosis. New data on the role of endolysosomes in autophagy (Madrigal-Matute et al., 2022; Zhang et al., 2014), atherosclerosis (Zhang et al., 2014), vesicle trafficking (chapter "Regulation of exosome release by lysosomal acid ceramidase in coronary arterial endothelial cells: Role of TRPML1 channel" by

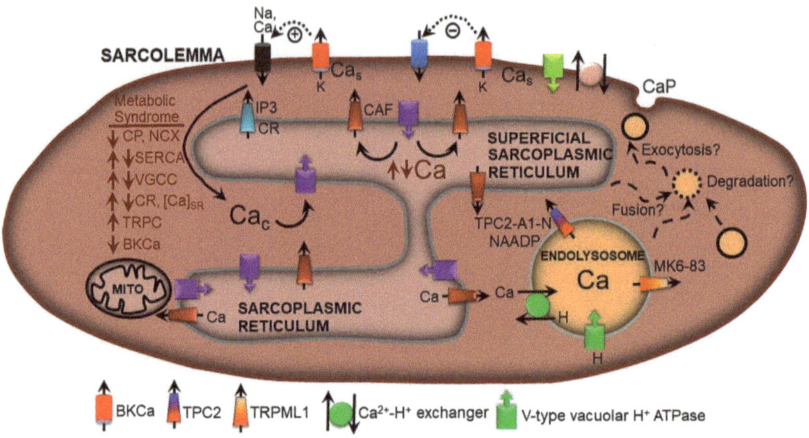

Fig. 12 Multiphasic model of coronary smooth muscle Ca^{2+} transporter regulation in atherosclerosis. See text for detailed explanation. Ca^{2+} transporters that functionally change in metabolic syndrome are listed on the left side of the cell. Note that the Ca^{2+} extrusion mechanisms are not technically decreased by metabolic syndrome, but plasmalemmal Ca^{2+} extrusion activity is already much slower in Ossabaw pigs compared to other healthy pigs (Neeb et al., 2010; Witczak & Sturek, 2004; Witczak, Wamhoff, & Sturek, 2006). Symbols and abbreviations: Ion transporter symbols in Fig. 1 are used here, while new others are shown in bottom of the figure. ↑, increase; ↓, decrease; TRPC, transient receptor potential canonical; BKCa, "big" Ca^{2+}-dependent K^{+} channel; VGCC, voltage-gated Ca channel; Ca_s, subsarcolemmal Ca; CP, plasma membrane Ca^{2+} pump; NX, Na^{+}-Ca^{2+} exchanger; CaP, hydroxyapatite; Ca_c, cytosolic Ca^{2+}; S, SERCA; TRPML1, transient receptor potential mucolipin 1; MK6-83, TRPML1 activator; CH, Ca^{2+}-H^{+} exchange; V-ATPase, vacuolar type H^{+} ATPase; NAADP, nicotinic acid adenine dinucleotide phosphate; TPC2, two pore channel 2; TPC2-A1-N, TPC2 activator; MITO, mitochondria.

Li et al.), the long-standing work on the role of exosomes in extracellular vascular calcification (Kapustin et al., 2015; Kapustin & Shanahan, 2016), and new findings we will present here have prompted the description of the model as multiphasic (Fig. 12). In response to MetS atherogenic stimuli the ion transporters shown in Fig. 12 increase and decrease function during the progression of atherosclerosis. More Ca^{2+} regulatory events make the process multiphasic.

Imaging of endolysosomes with LysoTracker suggests in Fig. 13 that major cell characteristics determining endolysosome labeling in human CSM are the severity of atherosclerosis and a blebbed appearance of the cell. The right coronary artery (RCA) was relatively devoid of flow-limiting atherosclerosis, i.e., closer to "healthy." In acutely dispersed cells 90% had a smooth appearance, while just 10% had a blebbed surface discernable

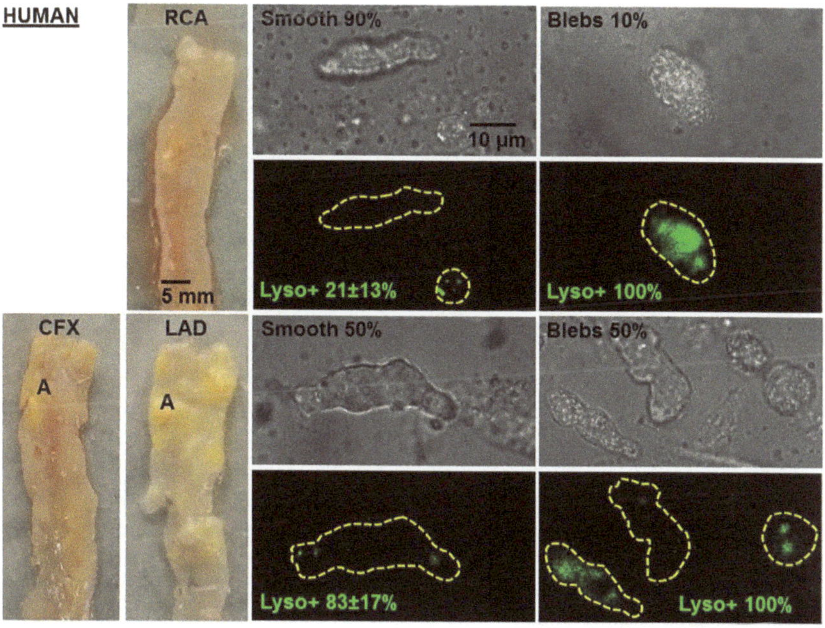

Fig. 13 Endolysosome imaging in human coronary smooth muscle cells in arteries with different severity of atherosclerosis and cell morphology. Arteries were harvested from an explanted heart that had ischemic cardiomyopathy. Arteries were cut longitudinally to create the strip. Atherosclerosis (A) was greatest in circumflex (CFX) and left anterior descending (LAD), as shown, for example, by the "A" label. Enzymatic dispersion provided cells for imaging endolysosomes with LysoTracker. Major cell characteristics that determined endolysosome labeling were the severity of atherosclerosis and a blebbed appearance of the cell. Number of cells analyzed: RCA = 10; CFX and LAD combined = 14.

in transmitted light microscopy. Endolysosomes were present in 100% of the blebbed RCA cells under widefield epifluorescence microscopy. In contrast, endolysosomes were present in just 21% of the smooth, "healthy" looking cells and the spots were quite punctate. The circumflex (CFX) and left anterior descending (LAD), in contrast, had significant atherosclerosis (A) with example spots labeled "A." Endolysosomes were plentiful in the entire cell population, as the smooth cells having endolysosomes increased to 83%. One limitation in this experiment is the limited sensitivity of the widefield epifluorescence imaging microscope, but the relative differences are still noteworthy.

Evidence for functional endolysosome Ca^{2+} regulation is shown by the Ca^{2+} release elicited by two pore channel 2 (TPC2) agonist in Fig. 14.

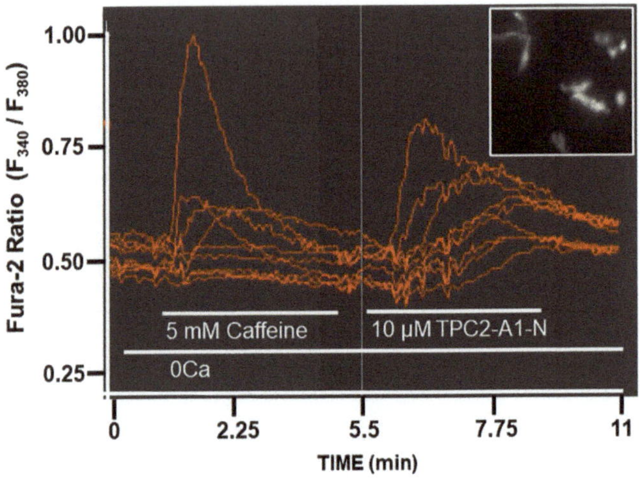

Fig. 14 Sarcoplasmic reticulum and endolysosome Ca^{2+} stores in coronary smooth muscle. Smooth muscle cells were acutely dispersed from a healthy lean Ossabaw swine for fura-2 imaging. Nine cells (shown in inset) were imaged simultaneously. Caffeine elicited SR Ca^{2+} release through the ryanodine receptor in 4 of 9 cells. TPC2-A1-N elicited endolysosome Ca^{2+} release through the TPC2 channel in 9 of 9 cells and the increased $[Ca^{2+}]_I$ was sustained longer than for caffeine.

Caffeine elicited SR Ca^{2+} store release through the ryanodine receptor in 4 of 9 cells. In the same cells the agonist TPC2-A1-N elicited endolysosome Ca^{2+} release through the TPC2 channel in 9 of 9 cells and the increased $[Ca^{2+}]_I$ was sustained longer than for caffeine. The consecutive stimulation of the cells in Ca^{2+}-free extracellular solution argues strongly for release from separate organelles. To our knowledge these are the first data for the endolysosomal Ca^{2+} store in native coronary smooth muscle cells. The selective agonist MK6-83 will enable determination of the role of the TRPML1 channel in native CSM. Cross-talk between the SR and endo/lysosome in CSM would seem to be inevitable. Several reports have shown in cultured pulmonary artery smooth muscle that receptor agonists (e.g., endothelin) induce contraction by Ca^{2+} release from endolysosomes that triggers Ca^{2+}-induced Ca^{2+} release from ryanodine-sensitive Ca^{2+} stores (Jiang et al., 2013; Negri, Faris, & Moccia, 2021). A follow-up study showed that the cultured cells exposed to hypoxia for 24–48 h have TPC2 channels upregulated and increased cell proliferation, which were blocked by the TPC2 channel inhibitor Ned-19 (Jiang et al., 2018). Impairment of endo-lysosomal Ca^{2+} responses and dynein-mediated autophagosome trafficking increased collagen deposition, thereby promoting smooth muscle remodeling

(Zhang et al., 2014). Further, arterial muscle calcification has been shown due to increased vesicle release in cultured mouse CSM having impaired TRPML1-mediated Ca^{2+} release from endolysosomes (Bhat et al., 2020).

The endolysosome is almost certainly involved in native CSM. Membrane interactions due to close spatial association of SR and endolysosomal Ca^{2+} stores was shown by the transfer of Ca^{2+} from the lysosome to the SR and triggering of Ca^{2+}-induced Ca^{2+} release (Jiang et al., 2013; Negri et al., 2021). The reverse may also be true, as we show in Fig. 12. Due to the impaired SERCA function in late stages of atherosclerosis (summarized in Fig. 5), there is a slow leak of Ca^{2+} from the SR. Although the Ca^{2+} could be released from superficial SR microdomains and extruded by the plasmalemmal Ca^{2+} pump and Na^{+}- Ca^{2+} exchanger, the activity of these is impaired. The endolysosome may effectively buffer the SR Ca^{2+} leak via the Ca^{2+}-H^{+} exchanger powered by the acid gradient generated from the V-ATPase. Herein is the delicate balance of fusion, degradation, and exocytosis in vesicular trafficking, especially for extracellular vascular calcification, which is due to accumulation of Ca^{2+} and inorganic phosphate to form hydroxyapatite (abbreviated as CaP in Fig. 12). Early detection of coronary artery calcification has been imaged in vivo in animal models and humans using ^{18}F positron emission tomography computed tomography (^{18}F PET-CT) (McKenney-Drake et al., 2018; McKenney-Drake, Rodenbeck, et al., 2016; McKenney-Drake, Territo, et al., 2016). Fig. 5 schematically illustrates this imaging, which occurs in mild fatty streaks and intimal thickening, several stages before vascular calcification can be detected with CT as a "Ca score" (Høilund-Carlsen et al., 2022; McKenney-Drake et al., 2018).

9. Conclusions and future directions

It is hoped that this review has dispelled the dogma that vascular disease would involve cellular Ca^{2+} overload. In fact, early increases in intracellular Ca^{2+} stores fade to decreased stores in late stages of coronary atherosclerosis. This is at the cost of increasing extracellular calcification. Study of animal models with high fidelity mimicry of human atherosclerosis could better enlighten development of therapeutics. Researchers utilizing these animal models should pay close attention to differences in vascular bed and, particularly, macro- vs. microvascular function. There is not a cacophony, but an orchestration, of changes in smooth muscle Ca^{2+} transporters in atherosclerosis. Some ion transport changes trigger dysfunction, while others compensate in attempts to maintain cellular functions, which ultimately fail in the face of continued noxious stimuli of the atherogenic milieu. As presented in Chapter 3 and elsewhere, the orderly trafficking of vesicles resulting in fusion, degradation, and exocytosis are novel mechanisms in vascular health and disease. There is much optimism about targeting these pathways for therapeutics for vascular calcification (Gerndt, Krogsaeter, Patel, Bracher, & Grimm, 2020; Jin et al., 2020; Tintut, Honda, & Demer, 2021). Use of optimal in vitro and whole animal models of atherosclerosis and calcification will greatly facilitate work in the area that may better translate to the clinic (Sturek et al., 2020). Another method that is a window to clinical studies is molecular imaging of atherosclerosis, more specifically vascular calcification, with ^{18}F PET-CT. Shown schematically in Fig. 5, ^{18}F PET-CT can be a tremendous tool to study some of the earliest steps in vascular calcification that involve endolysosomes and vesicle trafficking (McKenney-Drake et al., 2018).

Acknowledgments

This work was supported by National Institutes of Health grant P30 DK097512. We thank Eliana John for an excellent literature search and annotations. Due to space limitations, we have not presented all articles in detail and we apologize for any outstanding work that we have not cited.

References

Alshammary, A. F., Alharbi, K. K., Alshehri, N. J., Vennu, V., & Ali Khan, I. (2021). Metabolic syndrome and coronary artery disease risk: A Meta-analysis of observational studies. *International Journal of Environmental Research and Public Health*, *18*(4), 1773. https://doi.org/10.3390/ijerph18041773.

Amberg, G. C., & Navedo, M. F. (2013). Calcium dynamics in vascular smooth muscle. *Microcirculation*, *20*(4), 281–289. https://doi.org/10.1111/micc.12046 [doi].

Badin, J. K., Eggenberger, C., Rodenbeck, S. D., Alloosh, M., Hashmi, Z. A., Wang, I.-W., et al. (2021). Intracellular Ca^{2+} dysregulation in coronary smooth muscle is similar in coronary disease of humans and Ossabaw miniature swine. *Journal of Cardiovascular Translational Research*, *15*, 167–178. https://doi.org/10.1007/s12265-021-10153-5.

Badin, J. K., Kole, A., Stivers, B., Progar, V., Pareddy, A., Alloosh, M., et al. (2018). Alloxan-induced diabetes exacerbates coronary atherosclerosis and calcification in Ossabaw miniature swine with metabolic syndrome. *Journal of Translational Medicine*, *16*(1), 58. https://doi.org/10.1186/s12967-018-1431-9.

Berra-Romani, R., Mazzocco-Spezzia, A., Pulina, M. V., & Golovina, V. A. (2008). Ca2+ handling is altered when arterial myocytes progress from a contractile to a proliferative phenotype in culture. *American Journal of Physiology. Cell Physiology*, *295*(3), C779–C790. https://doi.org/10.1152/ajpcell.00173.2008.

Bhat, O. M., Li, G., Yuan, X., Huang, D., Gulbins, E., Kukreja, R. C., et al. (2020). Arterial medial calcification through enhanced small extracellular vesicle release in smooth muscle-specific Asah1 gene knockout mice. *Scientific Reports*, *10*(1), 1645. https://doi.org/10.1038/s41598-020-58568-5.

Choy, J. S., Luo, T., Huo, Y., Wischgoll, T., Schultz, K., Teague, S. D., et al. (2015). Compensatory enlargement of Ossabaw miniature swine coronary arteries in diffuse atherosclerosis. *IJC Heart & Vasculature, 6*, 4–11.

Climent, B., Santiago, E., Sanchez, A., Munoz-Picos, M., Perez-Vizcaino, F., Garcia-Sacristan, A., et al. (2020). Metabolic syndrome inhibits store-operated Ca(2+) entry and calcium-induced calcium-release mechanism in coronary artery smooth muscle. *Biochemical Pharmacology, 182*, 114222. https://doi.org/10.1016/j.bcp.2020.114222.

Dineen, S. L., Neeb, Z. P., Obukhov, A. G., & Sturek, M. (2016). Transient receptor potential channels in metabolic syndrome-induced coronary artery disease. In I. Levitan, & A. M. Dopico (Eds.), *Vascular ion channels in physiology and disease* (pp. 381–396). New York, NY: Springer. Reprinted from: In File.

Edwards, J. M., Alloosh, M. A., Long, X. L., Dick, G. M., Lloyd, P. G., Mokelke, E. A., et al. (2008). Adenosine A_1 receptors in neointimal hyperplasia and in-stent stenosis in Ossabaw miniature swine. *Coronary Artery Disease, 19*(1), 27–31. https://doi.org/10.1097/MCA.0b013e3282f262b4.

Edwards, J. M., Neeb, Z. P., Alloosh, M. A., Long, X., Bratz, I. N., Peller, C. R., et al. (2010). Exercise training decreases store-operated Ca^{2+} entry associated with metabolic syndrome and coronary atherosclerosis. *Cardiovascular Research, 85*(3), 631–640. https://doi.org/10.1093/cvr/cvp308.

Ehara, S., Kobayashi, Y., Yoshiyama, M., Shimada, K., Shimada, Y., Fukuda, D., et al., & Spotty Calcification Typifies the Culprit Plaque in Patients With Acute Myocardial Infarction. (2004). Spotty calcification typifies the culprit plaque in patients with acute myocardial infarction. *Circulation, 110*(22), 3424–3429. https://doi.org/10.1161/01.CIR.0000148131.41425.E9.

Fernandez-Tenorio, M., Gonzalez-Rodriguez, P., Porras, C., Castellano, A., Moosmang, S., Hofmann, F., et al. (2010). Short communication: Genetic ablation of L-type Ca2+ channels abolishes depolarization-induced Ca2+ release in arterial smooth muscle. *Circulation Research, 106*(7), 1285–1289. https://doi.org/10.1161/CIRCRESAHA.109.213967.

Gerndt, S., Krogsaeter, E., Patel, S., Bracher, F., & Grimm, C. (2020). Discovery of lipophilic two-pore channel agonists. *The FEBS Journal, 287*(24), 5284–5293. https://doi.org/10.1111/febs.15432.

Henry, P. D., & Bentley, K. I. (1981). Suppression of atherogenesis in cholesterol-fed rabbit treated with nifedipine. *The Journal of Clinical Investigation, 68*(5), 1366–1369. https://doi.org/10.1172/jci110384.

Hill, B. J. F., Dixon, J. L., & Sturek, M. (2001). Effect of atorvastatin on intracellular calcium uptake in coronary smooth muscle cells from diabetic pigs fed an atherogenic diet. *Atherosclerosis, 159*, 117–124.

Hill, B. J., Price, E. M., Dixon, J. L., & Sturek, M. (2003). Increased calcium buffering in coronary smooth muscle cells from diabetic dyslipidemic pigs. *Atherosclerosis, 167*(1), 15–23. https://doi.org/10.1016/s0021-9150(02)00381-7.

Hille, B. (1992). *Ionic channels of excitable membranes*. Sunderland, MA: Sinauer Associates, Inc.

Høilund-Carlsen, P. F., Piri, R., Madsen, P. L., Revheim, M. E., Werner, T. J., Alavi, A., et al. (2022). Atherosclerosis burden in diabetes mellitus: Assessment by PET imaging. *International Journal of Molecular Sciences, 23*, 10268 (In press).

House, S. J., Potier, M., Bisaillon, J., Singer, H. A., & Trebak, M. (2008). The non-excitable smooth muscle: Calcium signaling and phenotypic switching during vascular disease. *Pflügers Archiv, 456*(5), 769–785. https://doi.org/10.1007/s00424-008-0491-8.

Irkle, A., Vesey, A. T., Lewis, D. Y., Skepper, J. N., Bird, J. L., Dweck, M. R., et al. (2015). Identifying active vascular microcalcification by (18)F-sodium fluoride positron emission tomography. *Nature Communications, 6*, 7495. https://doi.org/10.1038/ncomms8495.

Jackson, W. F. (2020). Ion channels and the regulation of myogenic tone in peripheral arterioles. *Current Topics in Membranes, 85,* 19–58. https://doi.org/10.1016/bs.ctm.2020.01.002.

Jiang, Y. L., Lin, A. H., Xia, Y., Lee, S., Paudel, O., Sun, H., et al. (2013). Nicotinic acid adenine dinucleotide phosphate (NAADP) activates global and heterogeneous local Ca2+ signals from NAADP- and ryanodine receptor-gated Ca2+ stores in pulmonary arterial myocytes. *Journal of Biological Chemistry, 288*(15), 10381–10394. https://doi.org/10.1074/jbc.M112.423053.

Jiang, Y., Zhou, Y., Peng, G., Tian, H., Pan, D., Liu, L., et al. (2018). Two-pore channels mediated receptor-operated Ca(2+) entry in pulmonary artery smooth muscle cells in response to hypoxia. *International Journal of Biochemistry and Cell Biology, 97,* 28–35. https://doi.org/10.1016/j.biocel.2018.01.012.

Jin, X., Zhang, Y., Alharbi, A., Hanbashi, A., Alhoshani, A., & Parrington, J. (2020). Targeting two-pore channels: Current progress and future challenges. *Trends in Pharmacological Sciences, 41*(8), 582–594. https://doi.org/10.1016/j.tips.2020.06.002.

Kang, S., Dahl, R., Hsieh, W., Shin, A. C., Zsebo, K. M., Buettner, C., et al. (2016). Small molecular allosteric activator of the Sarco/endoplasmic reticulum Ca2+-ATPase (SERCA) attenuates diabetes and metabolic disorders. *Journal of Biological Chemistry, 291,* 5185–5198. https://doi.org/10.1074/jbc.M115.705012.

Kapustin, A. N., Chatrou, M. L., Drozdov, I., Zheng, Y., Davidson, S. M., Soong, D., et al. (2015). Vascular smooth muscle cell calcification is mediated by regulated exosome secretion. *Circulation Research, 116*(8), 1312–1323. https://doi.org/10.1161/CIRCRESAHA.116.305012.

Kapustin, A. N., & Shanahan, C. M. (2016). Emerging roles for vascular smooth muscle cell exosomes in calcification and coagulation. *Journal of Physiology, 594,* 2905–2914. https://doi.org/10.1113/JP271340.

Kim, J. Y., Mun, H. S., Lee, B. K., Yoon, S. B., Choi, E. Y., Min, P. K., et al. (2010). Impact of metabolic syndrome and its individual components on the presence and severity of angiographic coronary artery disease. *Yonsei Medical Journal, 51*(5), 676–682. https://doi.org/10.3349/ymj.2010.51.5.676.

Klabunde, R. E. (2005). *Cellular structure and function. In cardiovascular physiology concepts* (2nd ed., pp. 41–58). Lippincott Williams & Wilkins.

Kudryavtseva, O., Aalkjaer, C., & Matchkov, V. V. (2013). Vascular smooth muscle cell phenotype is defined by Ca2+−dependent transcription factors. *The FEBS Journal, 280*(21), 5488–5499. https://doi.org/10.1111/febs.12414.

Kumar, B., Dreja, K., Shah, S. S., Cheong, A., Xu, S. Z., Sukumar, P., et al. (2006). Upregulated TRPC1 channel in vascular injury in vivo and its role in human neointimal hyperplasia. *Circulation Research, 98*(4), 557–563.

Lakka, H. M., Laaksonen, D. E., Lakka, T. A., Niskanen, L. K., Kumpusalo, E., Tuomilehto, J., et al. (2002). The metabolic syndrome and total and cardiovascular disease mortality in middle-aged men. *JAMA, 288*(21), 2709–2716.

Landsberg, J. W., & Yuan, J. X. (2004). Calcium and TRP channels in pulmonary vascular smooth muscle cell proliferation. *News in Physiological Sciences, 19,* 44–50.

Leung, F. P., Yung, L. M., Yao, X., Laher, I., & Huang, Y. (2008). Store-operated calcium entry in vascular smooth muscle. *British Journal of Pharmacology, 153*(5), 846–857. https://doi.org/10.1038/sj.bjp.0707455.

Li, W., Chen, X., Riley, A. M., Hiett, S. C., Temm, C. J., Beli, E., et al. (2017). Long-term spironolactone treatment reduces coronary TRPC expression, vasoconstriction, and atherosclerosis in metabolic syndrome pigs. *Basic Research in Cardiology, 112,* 54. https://doi.org/10.1007/s00395-017-0643-0.

Libby, P. (2021). The changing landscape of atherosclerosis. *Nature, 592*(7855), 524–533. https://doi.org/10.1038/s41586-021-03392-8.

Lundberg, M. S., Curto, K. A., Bilato, C., Monticone, R. E., & Crow, M. T. (1998). Regulation of vascular smooth muscle migration by mitogen-activated protein kinase and calcium/calmodulin-dependent protein kinase II signaling pathways. *Journal of Molecular and Cellular Cardiology, 30*(11), 2377–2389. https://doi.org/10.1006/jmcc.1998.0795.

MacLennan, D. H., Asahi, M., & Tupling, A. R. (2003). The regulation of SERCA-type pumps by phospholamban and sarcolipin. *Annals of the New York Academy of Sciences, 986*, 472–480. Retrieved from http://www.ncbi.nlm.nih.gov/pubmed/12763867.

Madrigal-Matute, J., de Bruijn, J., van Kuijk, K., Riascos-Bernal, D. F., Diaz, A., Tasset, I., et al. (2022). Protective role of chaperone-mediated autophagy against atherosclerosis. *Proceedings of the National Academy of Sciences of the United States of America, 119*(14), e2121133119. https://doi.org/10.1073/pnas.2121133119.

Matchkov, V. V., Kudryavtseva, O., & Aalkjaer, C. (2012). Intracellular ca(2)(+) signalling and phenotype of vascular smooth muscle cells. *Basic & Clinical Pharmacology & Toxicology, 110*(1), 42–48. https://doi.org/10.1111/j.1742-7843.2011.00818.x [doi].

McKenney-Drake, M. L., Moghbel, M. C., Paydary, K., Alloosh, M., Houshmand, S., Moe, S. M., et al. (2018). 18F-NaF and 18F-FDG as molecular probes in the evaluation of atherosclerosis. *European Journal of Nuclear Medicine and Molecular Imaging, 45*(12), 2190–2200. https://doi.org/10.1007/s00259-018-4078-0.

McKenney-Drake, M. L., Rodenbeck, S. D., Owen, M. K., Schultz, K. A., Alloosh, M., Tune, J. D., et al. (2016). Biphasic alterations in coronary smooth muscle Ca2+ regulation in a repeat cross-sectional study of coronary artery disease severity in metabolic syndrome. *Atherosclerosis, 249*, 1–9. https://doi.org/10.1016/j.atherosclerosis.2016.03.032.

McKenney-Drake, M. L., Territo, P. R., Salavati, A., Houshmand, S., Persohn, S., Liang, Y., et al. (2016). ^{18}F-NaF PET imaging of early coronary artery calcification. *JACC: Cardiovascular Imaging, 9*, 627–628. https://doi.org/10.1016/j.jcmg.2015.02.026.

Mokelke, E. A., Dietz, N. J., Eckman, D. M., Nelson, M. T., & Sturek, M. (2005). Diabetic dyslipidemia and exercise affect coronary tone and differential regulation of conduit and microvessel K^+ current. *American Journal of Physiology. Heart and Circulatory Physiology, 288*(3), H1233–H1241. https://doi.org/10.1152/ajpheart.00732.2004.

Moosmang, S., Schulla, V., Welling, A., Feil, R., Feil, S., Wegener, J. W., et al. (2003). Dominant role of smooth muscle L-type calcium channel Cav1.2 for blood pressure regulation. *The EMBO Journal, 22*(22), 6027–6034.

Neeb, Z. P., Edwards, J. M., Alloosh, M., Long, X., Mokelke, E. A., & Sturek, M. (2010). Metabolic syndrome and coronary artery disease in Ossabaw compared with Yucatan swine. *Comparative Medicine, 60*(4), 300–315. Retrieved from http://www.ncbi.nlm.nih.gov/pubmed/20819380.

Negri, S., Faris, P., & Moccia, F. (2021). Endolysosomal ca(2+) signaling in cardiovascular health and disease. *International Review of Cell and Molecular Biology, 363*, 203–269. https://doi.org/10.1016/bs.ircmb.2021.03.001.

Ng, L. C., O'Neill, K. G., French, D., Airey, J. A., Singer, C. A., Tian, H., et al. (2012). TRPC1 and Orai1 interact with STIM1 and mediate capacitative Ca2+ entry caused by acute hypoxia in mouse pulmonary arterial smooth muscle cells. *American Journal of Physiology - Cell Physiology, 303*(11), C1156–C1172.

Nishida, M., Kuwahara, K., Kozai, D., Sakaguchi, R., & Mori, Y. (2015). *TRP channels: Their function and potentiality as drug targets, Tokyo*. Japan: Springer.

Ong, H. L., de Souza, L. B., & Ambudkar, I. S. (2016). Role of TRPC channels in store-operated calcium entry. In J. A. Rosado (Ed.), *Calcium entry pathways in non-excitable cells* (pp. 87–109). Cham: Springer International Publishing.

Ottolini, M., & Sonkusare, S. K. (2021). The calcium signaling mechanisms in arterial smooth muscle and endothelial cells. *Comprehensive Physiology, 11*(2), 1831–1869. https://doi.org/10.1002/cphy.c200030.

Owens, G. K., Kumar, M. S., & Wamhoff, B. R. (2004). Molecular regulation of vascular smooth muscle cell differentiation in development and disease. *Physiological Reviews*, *84*(3), 767–801. https://doi.org/10.1152/physrev.00041.2003.

Pauly, R. R., Bilato, C., Sollott, S. J., Monticone, R., Kelly, P. T., Lakatta, E. G., et al. (1995). Role of calcium/calmodulin-dependent protein kinase II in the regulation of vascular smooth muscle cell migration. *Circulation*, *91*(4), 1107–1115. https://doi.org/10.1161/01.cir.91.4.1107.

Phillips-Eakley, A. K., McKenney-Drake, M. L., Bahls, M., Newcomer, S. C., Radcliffe, J. S., Wastney, M. E., et al. (2015). Effect of high-calcium diet on coronary artery disease in Ossabaw miniature swine with metabolic syndrome. *Journal of the American Heart Association*, *4*, e001620.

Roberts, C. K., Hevener, A. L., & Barnard, R. J. (2013). Metabolic syndrome and insulin resistance: Underlying causes and modification by exercise training. *Comprehensive Physiology*, *3*, 1–58.

Searls, Y. M., Loganathan, R., Smirnova, I. V., & Stehno-Bittel, L. (2010). Intracellular Ca2+ regulating proteins in vascular smooth muscle cells are altered with type 1 diabetes due to the direct effects of hyperglycemia. *Cardiovascular Diabetology*, *9*, 8. https://doi.org/10.1186/1475-2840-9-8.

Shen, J., Halenda, S. P., Sturek, M., & Wilden, P. A. (2005). Novel mitogenic effect of adenosine on coronary artery smooth muscle cells: Role for the A_1 adenosine receptor. *Circulation Research*, *96*, 982–990.

Sturek, M. (2011). Ca^{2+} regulatory mechanisms of exercise protection against coronary artery disease in metabolic syndrome and diabetes. *Journal of Applied Physiology*, *111*(2), 573–586. https://doi.org/10.1152/japplphysiol.00373.2011.

Sturek, M., Alloosh, M., & Sellke, F. W. (2020). Swine disease models for optimal vascular engineering. *Annual Review of Biomedical Engineering*, *22*, 25–49. https://doi.org/10.1146/annurev-bioeng-082919-053009.

Sturek, M., Kunda, K., & Hu, Q. (1992). Sarcoplasmic reticulum buffering of myoplasmic calcium in bovine coronary artery smooth muscle. *The Journal of Physiology*, *451*, 25–48. Retrieved from http://www.ncbi.nlm.nih.gov/pubmed/1403813.

Sturek, M., Mokelke, E. A., Sindermann, J. R., Adam, L. P., & March, K. L. (2006). Molecular and cellular physiology of differentiated vascular smooth muscle. In J. T. Willerson, J. N. Cohn, H. J. J. Wellens, & D. R. Holmes (Eds.), *Cardiovascular medicine* (3rd ed., pp. 1511–1523). London: Springer-Verlag. Reprinted from: In File.

Sturek, M., Tune, J. D., & Alloosh, M. (2015). Ossabaw Island miniature swine: Metabolic syndrome and cardiovascular assessment. In M. M. Swindle, & A. C. Smith (Eds.), *Swine in the laboratory: Surgery, anesthesia, imaging, and experimental techniques* (3rd ed., pp. 451–465). Boca Raton: CRC Press. Reprinted from: In File.

Takahashi, Y., Watanabe, H., Murakami, M., Ohba, T., Radovanovic, M., Ono, K., et al. (2007). Involvement of transient receptor potential canonical 1 (TRPC1) in angiotensin II-induced vascular smooth muscle cell hypertrophy. *Atherosclerosis*, *195*(2), 287–296.

Tintut, Y., Honda, H. M., & Demer, L. L. (2021). Biomolecules orchestrating cardiovascular calcification. *Biomolecules*, *11*(10), 1482. https://doi.org/10.3390/biom11101482.

Trebak, M., Zhang, W., Ruhle, B., Henkel, M. M., Gonzalez-Cobos, J. C., Motiani, R. K., et al. (2013). What role for store-operated Ca(2)(+) entry in muscle? *Microcirculation*, *20*(4), 330–336. https://doi.org/10.1111/micc.12042.

Underwood, F. B., Laughlin, M. H., & Sturek, M. (1994). Altered control of calcium in coronary smooth muscle cells by exercise training. *Medicine and Science in Sports and Exercise*, *26*(10), 1230–1238. Retrieved from http://www.ncbi.nlm.nih.gov/pubmed/7799766.

Wamhoff, B. R., Bowles, D. K., Dietz, N. J., Hu, Q., & Sturek, M. (2002). Exercise training attenuates coronary smooth muscle phenotypic modulation and nuclear Ca^{2+} signaling.

American Journal of Physiology. Heart and Circulatory Physiology, 283(6), H2397–H2410. https://doi.org/10.1152/ajpheart.00371.2001.

Wamhoff, B. R., Bowles, D. K., McDonald, O. G., Sinha, S., Somlyo, A. P., Somlyo, A. V., et al. (2004). L-type voltage-gated Ca2+ channels modulate expression of smooth muscle differentiation marker genes via a rho kinase/myocardin/SRF-dependent mechanism. *Circulation Research, 95*(4), 406–414. https://doi.org/10.1161/01.RES.0000138582.36921.9e.

Witczak, C. A., & Sturek, M. (2004). Exercise prevents diabetes-induced impairment in superficial buffer barrier in porcine coronary smooth muscle. *Journal of Applied Physiology, 96*(3), 1069–1079. https://doi.org/10.1152/japplphysiol.00460.2003.

Witczak, C. A., & Sturek, M. (2005). Training-induced sarcoplasmic reticulum Ca^{2+} unloading occurs without Ca^{2+} influx. *Medicine and Science in Sports and Exercise, 37*(7), 1119–1125. https://doi.org/10.1249/01.mss.0000170125.25749.4d.

Witczak, C. A., Wamhoff, B. R., & Sturek, M. (2006). Exercise training prevents Ca^{2+} dysregulation in coronary smooth muscle from diabetic dyslipidemic Yucatan swine. *Journal of Applied Physiology, 101*(3), 752–762. https://doi.org/10.1152/japplphysiol.00235.2006.

Zhang, W., Halligan, K. E., Zhang, X., Bisaillon, J. M., Gonzalez-Cobos, J. C., Motiani, R. K., et al. (2011). Orai1-mediated ICRAC is essential for Neointima formation after vascular injury/novelty and significance. *Circulation Research, 109*(5), 534–542.

Zhang, Y., Xu, M., Xia, M., Li, X., Boini, K. M., Wang, M., et al. (2014). Defective autophagosome trafficking contributes to impaired autophagic flux in coronary arterial myocytes lacking CD38 gene. *Cardiovascular Research, 102*(1), 68–78. https://doi.org/10.1093/cvr/cvu011.

Zhang, S. L., Yu, Y., Roos, J., Kozak, J. A., Deerinck, T. J., Ellisman, M. H., et al. (2005). STIM1 is a Ca2+ sensor that activates CRAC channels and migrates from the Ca2+ store to the plasma membrane. *Nature, 437*(7060), 902–905. https://doi.org/10.1038/nature04147.

CHAPTER SIX

Specificity of Ca^{2+}-activated K^+ channel modulation in atherosclerosis and aerobic exercise training

Eric A. Mokelke[a], Mouhamad Alloosh[b], and Michael Sturek[b,*]
[a]W.L. Gore & Associates, Inc., Flagstaff, AZ, United States
[b]Department of Anatomy, Cell Biology, & Physiology, Indiana University School of Medicine, Indianapolis, IN, United States
*Corresponding author: e-mail address: msturek@iu.edu

Contents

Abstract

Vascular smooth muscle cells express several isoforms of a number of classes of K^+ channels. Potassium channels play critical roles in the regulation of vascular smooth muscle contraction as well as vascular smooth muscle cell proliferation or phenotypic modulation. There is ample evidence that it is Ca^{2+} that enables these two seemingly disparate functions to be tightly coupled both in healthy and disease processes. Because of the central position that potassium channels have in vasocontraction, vasorelaxation, membrane potential, and smooth muscle cell proliferation, these channels continue to possess the potential to serve as novel therapeutic targets in cardiovascular disease. While there are questions that remain regarding the complete interactions between K^+ channels, vascular regulation, smooth muscle cell proliferation, and phenotypic modulation in physiological and pathophysiological conditions, a broad understanding of the contributions of each class of K^+ channel to contractile and proliferative states of the vasculature has been reached. This brief review will discuss the current understanding

Current Topics in Membranes, Volume 90
ISSN 1063-5823
https://doi.org/10.1016/bs.ctm.2022.09.005

of the role of K^+ channels in vascular smooth muscle cells in health and disease using the porcine vascular smooth muscle cell model with particular attention to new scientific discoveries contributed by the authors regarding the effect of endurance exercise on the function of the K^+ channels.

1. General information on main types of K^+ channels

The membrane potential of vascular smooth muscle (VSM) cells is regulated in part by several types of K^+ channels. Vessel diameter and the tone of the vasculature is known to be regulated in part by membrane potential of the smooth muscle cells. Therefore, the diameter of blood vessels as well as the tone of the vasculature are determined in part by the activity of various K^+ channels (Nelson & Quayle, 1995). The types of K^+ channels that have been studied the to the greatest extent in vascular smooth muscle are voltage-gated K channels (Kv), calcium-activated K channels (KCa), ATP-sensitive K^+ channels (KATP), and inward rectifier K^+ channels (Kir).

1.1 Voltage-gated K channels (Kv)

Nelson and Quayle have reported that voltage-gated K^+ channels (Kv) can be subdivided into 12 subfamilies (Kv1-Kv12). The Kv channels consists of 4 α-subunits and 4 β-subunits. The α-subunits forms the K^+ selective pore and contains 6 transmembrane domains (s1-s6) and a P loop between the s5 and s6 domains. The s1-s4 region composes the voltage sensing domain (Nelson & Quayle, 1995).

In response to membrane depolarization from resting potential, the Kv channels open allowing for movement of K^+ ions. Repolarization of the membrane occurs as a result of the opening of Kv channels, which allows efflux of K^+ ions, thus returning the membrane to its resting potential. As in cardiac myocytes (Snyders, 1999), this Kv channel function serves to provide a limit to membrane depolarization. Especially in vascular myocytes, the function of limiting membrane potential depolarization impacts the contraction of a vessel by attenuating the entry of Ca^{2+} ions through L-type voltage-gated Ca^{2+} channels. The interaction of the Kv channels and L-type Ca^{2+} channel activity defines a main action of Kv channels in maintaining vascular tone (Ko, Han, Jung, & Park, 2008; Korovkina & England, 2002; Nelson & Quayle, 1995).

1.2 ATP-sensitive K^+ channels (KATP)

ATP-sensitive K^+ channels are also expressed in cardiac as well as VSM cells. The mechanism of activation is dependent on the local concentration of ATP and/or ADP and thus considered to be tightly coupling to by-products of cellular metabolism. As a result of this sensitivity to ATP or ADP, the function of KATP channels can be considered to be protection against ischemia. As a result of metabolic changes, these channels can affect vascular tone and as a result blood flow (Nelson, 1993).

The KATP channel is made up of four pore-forming Kir6 subunits (Kir6.1 or Kir6.2) and four regulatory sulfonylurea (SUR) receptor subunits. It is known that the KATP channel in VSM is formed by a combination of the Kir6.1 and SUR2B subunits. It has been shown that Kir6.1 subunit forms the pore and has a crucial role in the regulation of vascular tone (Dogan, Yildiz, Arslan, & Ulusoy, 2019).

The opening of KATP results in membrane hyperpolarization, and due to the closing of voltage-gated Ca^{2+} channels, internal Ca^{2+} concentrations are reduced, resulting in relaxation of the blood vessel (Aguilar-Bryan et al., 1998; Nelson, Patlak, Worley, & Standen, 1990; Standen et al., 1989).

Since KATP channels are sensitive to both ATP and ADP, the activation of KATP channels could occur via in reduction in intracellular ATP and could also be activated by elevation in intracellular ADP. Other activators of KATP channels may be intracellular acidification that occurs as a result of hypoxia or an actual oxygen-sensing mechanism (Daut et al., 1990; von Beckerath, Cyrys, Dischner, & Daut, 1991). Berne et al. have mentioned that ischemia causes the release of adenosine from cardiac myocytes and since adenosine is a potent dilator of coronary arteries it has been proposed that adenosine may act as a metabolic vasodilator in the coronary circulation (Berne, 1980).

1.3 Inward rectifier K^+ channels (Kir)

Inward rectifier K^+ channels can participate in membrane hyperpolarization when extracellular K^+ concentration becomes elevated. The inward K^+ current is greater than the outward K+ current when Kir channels open. Local K^+ levels can be elevated when muscle contraction occurs. The inward K+ current resulting when these channels open is greater than the outward K current (Caesar, Akgoren, Mathiesen, & Lauritzen, 1999). Inward rectifier K^+ channels consist of tetramers where each subunit has two transmembrane domains. When extracellular K^+ concentration is elevated, Kir channels are activated and thereby causes membrane hyperpolarization toward the K^+ equilibrium potential (Martens & Gelband, 1998; Sorensen, Braunstein, Holstein-Rathlou, & Salomonsson, 2012). Of the large gene family of Kir subfamilies (Kir1.x to Kir7.x), the most relevant Kir expressed in the vasculature belong to the Kir2.x (strong inward rectifier) and Kir6.x (weakly inward rectifier) subfamilies.

1.4 Calcium-activated K channels (BKCa)

Large conductance K^+ channels that are activated by intracellular Ca^{2+} and membrane depolarization have been found in virtually every type of smooth muscle. BKCa channels are activated by local intracellular Ca^{2+} concentrations and these channels are inhibited by certain compounds applied externally, such as charybdotoxin, tetraethylammonium, and iberiotoxin (Benham, Bolton, Lang, & Takewaki, 1986; Brayden & Nelson, 1992; Giangiacomo, Garcia, & McManus, 1992; Langton, Nelson, Huang, & Standen, 1991; Miller, Moczydlowski, Latorre, & Phillips, 1985; Nelson, 1993). Upon activation, these K^+ channels permit K^+ to pass readily through them and thus have often been called "big" KCa, channels or "maxi" KCa, channels (Nelson & Quayle, 1995). Similar to Kv channel construction, BKCa, channels have the six hydrophobic segments (Sl-S6) that comprise the transmembrane domain, the pore and the voltage sensor. Unique and necessary to this channel is a Ca^{2+} sensing segment (S7-S10). The BKCa channel family contains channels with different conductance and sensitivity properties which may be a result of a single mammalian gene responsible for generating many variant channel peptides (Inoue, Kitamura, & Kuriyama, 1985; Toro, Vaca, & Stefani, 1991).

2. Physiological roles of KCa channels

KCa channels play an important role in the control of myogenic tone. It has been proposed that pressure-induced membrane depolarization and increases in intracellular Ca^{2+} activate KCa channels (Brayden & Nelson, 1992; Nelson, 1993). Due to the depolarization of the membrane, when the KCa channel is activated by the local increases in Ca^{2+} levels, K^+ efflux causes a shift from membrane depolarization to a return of the membrane potential toward its resting value (Jackson, 2000, 2005; Ko et al., 2008; Park et al., 2007; Tanaka, Koike, & Toro, 2004; Waldron & Cole, 1999). A similar sequence of events is reported to occur with chemical-induced depolarization and vasoconstriction. Fig. 1 represents many of the relationships between

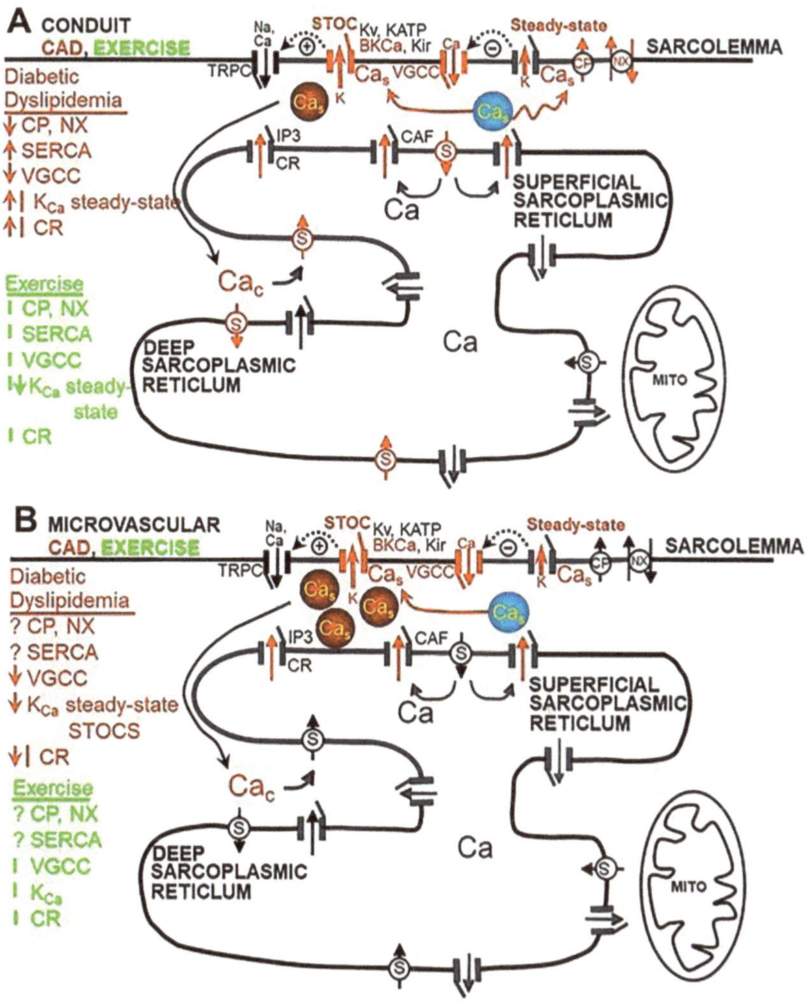

Fig. 1 Schematic model of Ca^{2+} regulation abnormalities in CSM in diabetic dyslipidemia and attenuation of abnormalities by exercise training. See also a more comprehensive review of exercise training effects on coronary conduit artery smooth muscle (Sturek, 2011). (A) Conduit, CAD, Exercise. CSM cells from a conduit artery of a sedentary pig show abnormalities of Ca^{2+} regulatory processes, proteins, and membrane domains that influence Ca^{2+} signaling. The red font and structures denote functional and molecular changes (actual protein expression or membrane domain differences). For example, SERCA function and SERCA2b protein (S) are increased in diabetic dyslipidemia. There is a retraction of the superficial SR from the sarcolemma, thereby altering the coupling of SR with sarcolemmal ion transporters. Overall, decreased Ca^{2+} extrusion via the plasmalemmal Ca^{2+} pump and Na^{+}/Ca^{2+} exchange is a major defect in Ca^{2+} regulation. SERCA and KCa channel function typically increase, including a closer coupling of Ca^{2+} release from caffeine-sensitive SR Ca^{2+} release channels and increased subsarcolemmal Ca^{2+} (Ca_s), perhaps partially compensating for other defects. We speculate that subsarcolemmal Ca^{2+} (Ca_s; blue sphere) is tonically increased

(Continued)

KCa channels and intracellular and subsarcolemmal Ca^{2+} and K^{+} channel interactions in sedentary (healthy), in pathophysiological conditions (diabetic dyslipidemia and coronary artery disease [CAD]), and the effects of chronic exercise in smooth muscle cells from conduit and microvessels. Note that while this is a simplified representation, the integrative nature of K^{+} channels and the various Ca^{2+} depots demonstrate the elegant importance of K^{+} channels in the vasculature. Further, longer durations of diabetic dyslipidemia or metabolic syndrome can elicit even more severe coronary disease and ion transport dysfunction as presented in Chapter 5 and Chapter 8 in this volume.

Attention is now put on the aforementioned local increases in intracellular Ca^{2+} and the relationship to KCa channels. Seminal work from Nelson's group described a tight relationship under normal conditions of ryanodine-sensitive Ca^{2+} channels to activation of KCa channels in smooth

Fig. 1—Cont'd due to the impaired coupling of slow SR Ca^{2+} release from ryanodine and caffeine (CAF)-sensitive Ca^{2+} release channels with Ca^{2+} extrusion by the plasmalemmal Ca^{2+} pump (CP) and Na^{+}-Ca^{2+} exchanger (NX). This elevation in subsarcolemmal Ca^{2+} (blue sphere) could increase steady-state BKCa current in diabetic dyslipidemia. Rare Ca^{2+} sparks provide higher Ca^{2+} concentration (red sphere) in the subsarcolemmal space near BKCa channels to elicit rare STOC events in conduit CSM. [The IP3-sensitive and caffeine/ryanodine-sensitive Ca^{2+} release channels (CR) are shown generically as one channel for simplicity, but are not the same molecular entity.] Voltage-gated Ca^{2+} channel current is downregulated. The net effect of these Ca^{2+} regulation abnormalities is increased cytosolic Ca^{2+} (↑ Ca_c). The precise subcellular localization of Ca^{2+} transport molecules and processes is emphasized for the crucial role in CSM function. Aerobic exercise training attenuates these abnormalities in Ca^{2+} transport protein and membrane domains. The green font denotes those changed by exercise training. For example, the vertical green bar at CP, NCX shows that the decreased (↓) CP, NX in diabetic dyslipidemia was returned to normal by exercise. The red vertical bar and ↑ at KCa indicates no change or an increase in diabetic dyslipidemia vs. healthy control. (B) Microvascular, CAD, Exercise. Main differences from conduit CSM are the profoundly greater subsarcolemmal Ca^{2+} shown by the 3 red spheres indicative of more Ca^{2+} sparks and almost 10-fold more frequent STOC events compared to conduit CSM. The effects of diabetic dyslipidemia and exercise on CP, NX, and SERCA have not been studied (?). Abbreviations and symbols: TRPC, transient receptor potential canonical channels; VGCC, voltage-gated Ca^{2+} channels; CP, plasmalemma Ca^{2+} ATPase (PMCA pump); NX, Na^{+}-Ca^{2+} exchange; K_{Ca}, Ca^{2+}-activated K channels; Ca_s, subsarcolemmal Ca^{2+}; Mito, mitochondria; IP3, inositol trisphosphate; CR, ryanodine-sensitive Ca^{2+} release channels; Ca_c, cytosolic Ca^{2+}; S = SERCA = sarcoplasmic-endoplasmic reticulum Ca^{2+} ATPase; SR, sarcoplasmic reticulum; +, hyperpolarization-induced stimulation of ion flux; −, hyperpolarization-induced inhibition of channel opening; ↑, increased; ↓, decreased; |, no changes in diabetic dyslipidemia or attenuated or returned diabetic dyslipidemia-induced changes toward healthy my chronic exercise.

muscle cells. The significance of this data is that for the first time, local release events at the ryanodine receptor were seen to cause opening of KCa channels resulting in efflux of K. The coupling of ryanodine receptors and KCa channels appeared to demonstrate the central role of ryanodine-sensitive Ca^{2+} channels in regulating muscle activation by the opening of KCa channels (Nelson et al., 1995). Using confocal microscopy, several groups have shown the occurrence of the small releases of Ca^{2+} near the cell membrane and the ryanodine sensitive receptors and because of their transient and quantal nature, referred to them as Ca^{2+} sparks. Further data reinforced the notion that local Ca^{2+} release events, sparks, are indeed the trigger to open sarcolemmal KCa channels in a consequential fashion because of the appearance of sparks immediately below the sarcolemma (Brayden & Nelson, 1992; Knot & Nelson, 1995; Nelson et al., 1990; Nelson & Quayle, 1995).

Patch clamp studies on isolated smooth muscle cells provide more evidence that when sparks are generated, outward current transients were observed that with similar time course to the appearance of the sparks. Furthermore, these outward currents could be inhibited by ryanodine or by depletion of the SR Ca^{2+} store with the sarco-endoplasmic reticulum Ca^{2+}-ATPase inhibitor thapsigargin. Finally, the outward currents were inhibited by tetraethylammonium and iberiotoxin, providing strong evidence that the Ca^{2+} spark-induced outward currents were a result of activation of KCa channels. Taken together, the functional coupling of ryanodine-sensitive Ca^{2+} channels and KCa channels appear to be mediated by Ca^{2+} sparks with the resultant outward current providing a means to prevent further depolarization of membrane. At the intact vessel level, as intracellular Ca^{2+} increases and intravascular pressure is increased, activation of outward currents via the KCa channels provides a mechanism for relaxation of the vessel (Jaggar et al., 1998; Somlyo & Somlyo, 1994). Calcium sparks have been subsequently measured in myocytes from portal vein (Mironneau, Arnaudeau, Macrez-Lepretre, & Boittin, 1996) and smooth muscle from numerous vascular beds, airway (Sieck, Kannan, & Prakash, 1997), esophagus (Kirber et al., 2001), and ileum (Gordienko, Bolton, & Cannell, 1998).

The outward currents resulting from the Ca^{2+} sparks, due to their transient nature, have been coined Spontaneous Transient Outward Currents (STOCs) by others and can be used as an indirect measure of spark activity (Brayden & Nelson, 1992; Knot & Nelson, 1998; Knot, Standen, & Nelson, 1998).

3. Effects of chronic exercise on reduction of cardiovascular disease and underlying K^+ channel adaptations

The benefits of physical activity and exercise on all body systems are well described. They include an increase in aerobic capacity and attenuation of coronary artery disease, which were recently reviewed (Sanchis-Gomar et al., 2022). One focus of our laboratory has been to develop and test the hypothesis that because K^+ channels play an important role in the regulation of blood flow, the pathophysiological process of cardiovascular disease will negatively impact K^+ channel function and, conversely, chronic exercise will mitigate this adverse effect on the K^+ channels. Evidence that chronic exercise enhances circulation through a mechanism involving K^+ channels of smooth muscle cells was published by Bowles, Laughlin, and Sturek, (1998). This study demonstrated that in conduit vessels of the coronary circulation of healthy swine, exercise training increased the K^+ channel contribution to regulation of coronary arterial tone (Bowles et al., 1998). The complexity of evaluating the singular effect of disease or exercise on K^+ channels is appreciated when one understands that within the coronary arterial tree the contribution of specific K channels to regulation of tone can be dependent on the relative vessel size and/or function, i.e., conduit artery vs. resistance artery. (Bowles et al., 1998). Specifically for KCa channels, Dick and Tune showed that exercise training enhances the contribution of KCa channel activation to exercise induced coronary vasodilation (Dick & Tune, 2010). Significant effects on hemodynamic variables as a result of chronic exercise that are based on changes in BKCa channels also occur in other arterial beds, thus the alterations in ionic channels are not isolated to the coronary circulation (Shi, Zhao, Zeng, Li, & Liu, 2012).

The important role that K^+ channels play in blood flow control in healthy animals emphasizes the potential consequences of alterations in the contribution of K^+ channels as a result of disease or physical activity and exercise. In chronically awake swine, blockade of KCa channels with tetraethylammonium results in mild coronary vasoconstriction at rest that progressively increases with increasing exercise intensity. In contrast, blocking KATP channels results in pronounced coronary vasoconstriction at rest, that decreases with increasing exercise intensity (Merkus, Sorop, Houweling, Hoogteijling, & Duncker, 2006). Despite the mechanistic explanation of the benefits of physical activity and exercise on some hemodynamic properties, the authors caution against a simple superimposition of this explanation onto the benefits of exercise on the background of a diseased state or aging. There are data that demonstrate that pathophysiological conditions can alter K^+ channels. With respect to K^+ channels and vasodilation, the effects of aging can be positively affected by physical activity and exercise. Exercise can reverse age related declines in mesenteric and coronary vasodilatory properties by attenuating BKCa channel downregulation (Albarwani, Al-Siyabi, Baomar, & Hassan, 2010; Shi et al., 2012).

Potassium channel current density is impacted by hypercholesterolemia in coronary artery smooth muscle cells, however the effect is dependent on the relative location of the coronary artery, i.e., conduit or microvessel. Hypercholesterolemia impairs coronary vasodilation or coronary flow mediated by K^+ channels (Hasdai et al., 1999; Mokelke, Dietz, Eckman, Nelson, & Sturek, 2005). Fig. 2 represents summary data from studies evaluating coronary blood flow (CBF). Coronary blood flow was measured using a coronary flow wire in control, diabetic dyslipidemic and diabetic dyslipidemic exercise trained pigs. Endurance exercise training reversed the diabetic dyslipidemia-induced reduction in coronary blood flow. This was the first time that the hypothesis that exercise training would mitigate the undesirable effects of diabetic dyslipidemia on coronary blood flow was

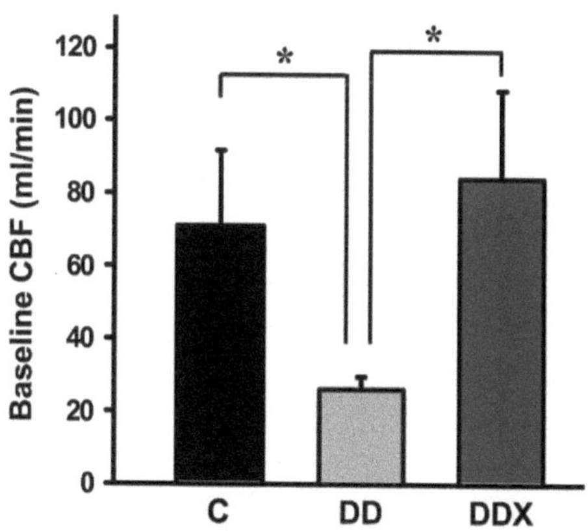

Fig. 2 Baseline coronary blood flow (CBF) was significantly depressed in diabetic dyslipidemic (DD) animals. Exercise training prevented the impairment in CBF. $n = 6$ for control (C), DD, and treadmill-trained animals. *Significantly different from DD, $P < 0.05$. *Reproduced with permission from Mokelke, E. A., Dietz, N. J., Eckman, D. M., Nelson, M. T., & Sturek, M. (2005). Diabetic dyslipidemia and exercise affect coronary tone and differential regulation of conduit and microvessel K^+ current.* Am J Physiol Heart Circ Physiol, 288*(3), H1233-H1241. https://doi.org/10.1152/ajpheart.00732.2004.*

tested. The data clearly show that endurance exercise training reversed the reduction in CBF caused by diabetic dyslipidemia. Experimental coronary occlusion also modulates steady-state K^+ channel current in coronary arteriole smooth muscle cells (Heaps, Tharp, & Bowles, 2005), as reviewed in Chapter 7 of this volume. However, in smooth muscle cells isolated from conduit arteries, a hypercholesterolemic diabetic state did not show a difference in steady-state K^+ channel current (Mokelke et al., 2005). The contribution of Kv channels to adenosine-mediated dilation of pig coronary arterioles is abolished by hypercholesterolemia (Heaps et al., 2005). Thus, a subtype of K channels that are not sensitive to Ca^{2+} are differentially affected by hypercholesterolemia.

When the effect of exercise on steady-state BKCa currents are quantified, atherogenic diet as well as diabetes and diabetic dyslipidemia has been reported to increase steady-state BKCa current in conduit coronary artery smooth muscle cells, an effect that was mitigated by exercise (Mokelke et al., 2003). In contrast, Fig. 3 shows a decrease in steady-state BKCa current in microvessel coronary artery smooth muscle cells (Mokelke et al., 2005). Based on similar differential results with K^+ currents, it should not be a surprise to see a differential effect of a disease on BKCa currents. Additionally, another measure of BKCa activity is the quantity of spontaneous transient outward current (STOC) events. Because Ca^{2+} sparks activate STOCs, STOC events can be used as a bioassay for Ca^{2+} sparks. STOCs occurred at a very low frequency in conduit artery smooth muscle cells, however, in microvessel smooth muscle cells, the STOC occurrence was an order of magnitude greater. The Ca^{2+} spark frequency in microvessels exhibited a similar pattern where spark frequency was significantly reduced in smooth muscle cells obtained from microvessels, but this reduction was mitigated by exercise training. Taken together, in a pathophysiological condition such as diabetic dyslipidemia, where adenosine-induced coronary blood flow is reduced and rescued by a program of exercise training, the changes in STOC occurrence and spark frequency provides a mechanistic explanation for the improvement in coronary blood flow. Fig. 3 underscores the differences between smooth muscle cells obtained from microvessel arteries and those obtained from conduit arteries. Spontaneous transient outward currents (STOCs) were decreased to just 15% of control in smooth muscle cells from microvessels obtained from DD animals, but not in cells from conduit arteries. The data clearly demonstrate not only the differences

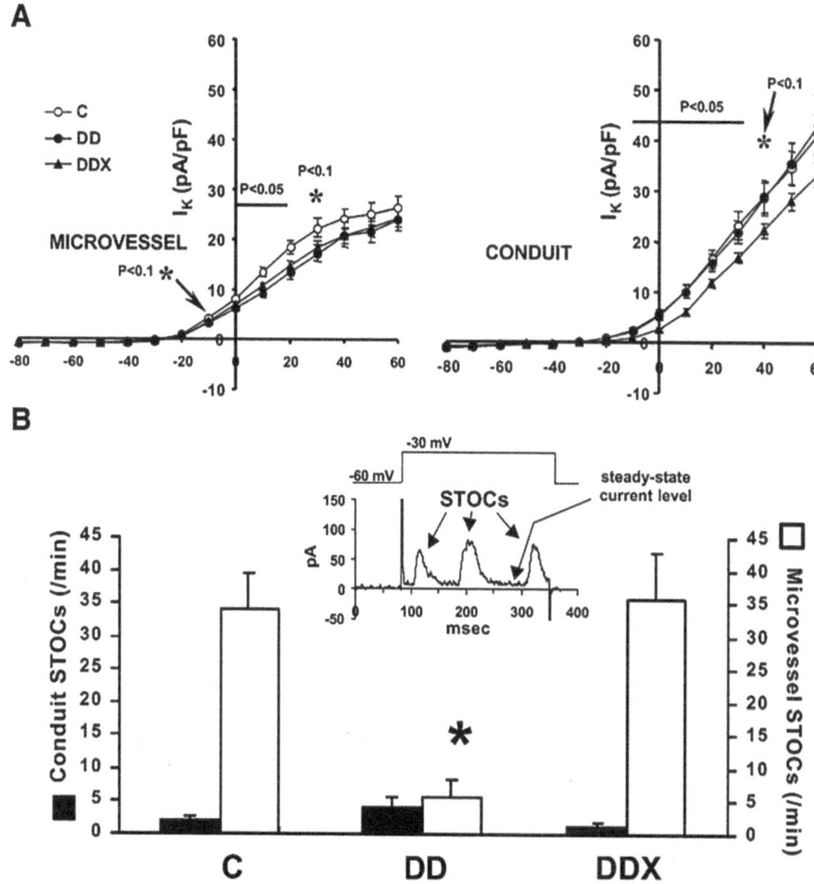

Fig. 3 Spontaneous transient outward currents (STOCs) were attenuated in smooth muscle cells from microvessels obtained from DD animals but not in cells from conduit arteries. A: whole cell steady-state K^+ current (I_K) density-voltage relations resulting from step depolarizations from −70 to +60 mV from a holding potential of −80 mV in cells from microvascular (left) and conduit (right) coronary smooth muscle cells. Steady-state current density was generally decreased in DD ($n = 19$) and DDX ($n = 34$) versus C ($n = 34$). Pairwise differences occurred at the $P < 0.05$ level in C vs. DD (0 mV to +20 mV) and DDX (10 mV) and at the $P < 0.1$ level in C vs. DDX at +20 mV. Conduit cell current densities were decreased in DDX ($n = 42$) versus C ($n = 24$) from −10 to +40 mV, whereas DDX was less than DD ($n = 33$) at voltages greater than −20 mV. Pairwise differences occurred at the $P < 0.05$ level in DDX vs. DD and C (0 mV and +10 mV) and at the $P < 0.1$ level in DDX vs. DD and C at −10 mV and +20 mV. B: inset illustrates that step depolarization from a holding potential of −60 mV to −30 mV for 270 ms results in a small steady-state potassium current, frequently accompanied by STOCs (3 in this case). STOC frequency in smooth muscle cells from coronary microvessels of DD ($n = 19$) was significantly lower than STOC frequency in C ($n = 35$) and DDX ($n = 34$) (*$P < 0.05$). STOC frequency in cells from conduit vessels, which exhibit low frequency of events when compared with control microvessels, was not different across groups. *Reproduced with permission from Mokelke, E. A., Dietz, N. J., Eckman, D. M., Nelson, M. T., & Sturek, M. (2005). Diabetic dyslipidemia and exercise affect coronary tone and differential regulation of conduit and microvessel K^+ current.* Am J Physiol Heart Circ Physiol, *288(3), H1233-H1241. https://doi.org/10.1152/ajpheart.00732.2004.*

in the contribution of STOCs to Ca^{2+} regulation/homeostasis, but also the difference in the K^+ current vs. voltage relationship between cells obtained from coronary artery microvessels and conduit arteries. One possible explanation for the large number of STOCs in microvessels in the cells obtained from diabetic dyslipidemic exercise trained pigs is that STOCs are an indication that the cells are able to eliminate excess or unwanted Ca^{2+} from various Ca^{2+} depots.

A more difficult result to explain is the low STOC occurrence in smooth muscle cells obtained from conduit arteries. In a very similar model of diabetic dyslipidemia and exercise, we have reported an increased coupling between Ca^{2+} release and KCa channel activation in conduit coronary artery in diabetic dyslipidemia (Mokelke et al., 2003). In coronary artery smooth muscle cells loaded with the calcium indicator fura-2, we reported no difference in caffeine-induced maximal intracellular Ca^{2+} concentrations between any diseased group compared to the control or exercise trained group (Fig. 4). The maximal BKCa current elicited by the maximal caffeine-induced intracellular Ca^{2+} transient, however, was significantly greater in all diseased groups compared to the control and the exercise trained group. A greater caffeine-sensitive current could be the result of a greater number of KCa channels, which is possible due to the modestly greater steady-state K^+ currents. The more likely interpretation is an alteration in the coupling of the SR Ca^{2+} release channels to the KCa channels such that the same amount of Ca^{2+} released from the SR activated more KCa channels. It is unlikely that the size of the caffeine-sensitive Ca^{2+} stores was greater in the diabetic dyslipidemic group because there were no differences between groups in the peak fura-2 ratios on caffeine application. Fig. 4 represents data from caffeine-induced K currents obtained from electrophysiological studies as described in the legend. Because the immunohistochemistry and immunoblotting showed no statistical differences in BKCa channel protein between groups, this rules out the possibility that there were a greater number of KCa channel subunits in the groups that exhibited the greatest caffeine-induced outward current. A more likely explanation is that the function, including functional coupling of SR Ca^{2+} release to the BKCa channels, was increased (Mokelke et al., 2003). Nevertheless, it is notable in Fig. 3 that microvascular coronary smooth muscle cell regulation appears to have a profoundly large component of Ca^{2+} sparks and STOCs (Mokelke et al., 2005).

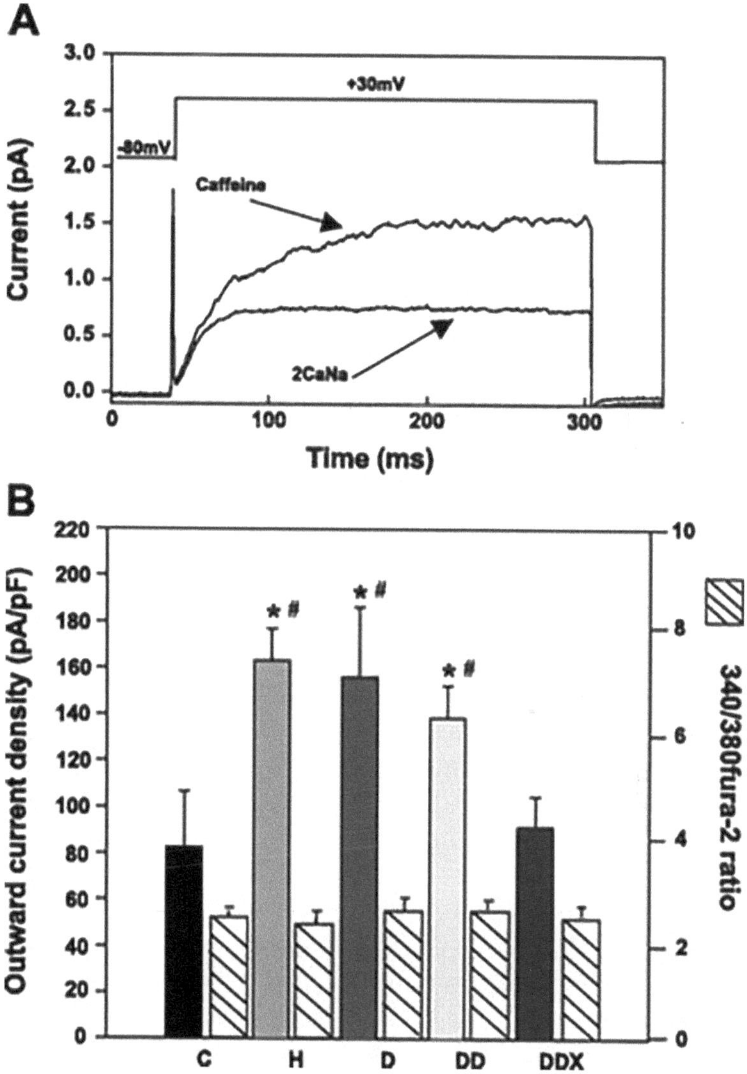

Fig. 4 Ca^{2+}-sensitive K (KCa) channel current is elevated in hyperlipidemia, diabetes, and diabetic dyslipidemia. *A*: raw current tracings in a Ca^{2+}-containing solution (2CaNa) and with addition of 5 mM caffeine in the superfusate. *A* also indicates voltage step protocol. *B*: H, D, and DD treatments resulted in a significant increase in caffeine-induced outward current compared with C and DDX despite similar cytoplasmic Ca^{2+} concentrations. Application of caffeine (5 mM) activated KCa channels by unloading a majority of the Ca^{2+} sequestered in the sarcoplasmic reticulum. *Peak caffeine-induced I_K density was greater in H, D, and DD compared with C; $^{\#}$peak current in H, D, and DD was greater than in DDX. There were no differences in the fura 2 ratios between groups, $P < 0.05$. *Reproduced with permission from Mokelke, E. A., Hu, Q., Song, M., Toro, L., Reddy, H. K., & Sturek, M. (2003). Altered functional coupling of coronary K^+ channels in diabetic dyslipidemic pigs is prevented by exercise.* Journal of Applied Physiology, 95*(3), 1179–1193. https://doi.org/10.1152/japplphysiol.00972.2002.*

4. Conclusions and future directions

The pathophysiological conditions that were imposed on the swine model involved clinically relevant diseases of diabetes, diabetic dyslipidemia, and hyperlipidemia. The therapy that was superimposed on some of these conditions was aerobic exercise. The novel findings indicate that there are alterations in the expression or function of the channels as a result of a pathphysiological condition, in addition to the results of functional coupling alterations due to the disease states reveal a potential new understanding of the function of BKCa channels. Furthermore, the authors have shown some evidence that introduces the notion that increased steady-state KCa in conduit coronary smooth muscle cells in diabetes, hyperlipidemia, and diabetic dyslipidemia may be an attempt to compensate for increased Ca influx and cycling. Finally, the authors would like to highlight that the differential regulation of KCa channels in conduit vs. microvascular smooth muscle compounds the complexity in making interpretations regarding the effects of coronary artery disease induced by diabetes and diabetic dyslipidemia and the mitigating effects of exercise. It is unclear from these data whether the increase in BKCa current due to the disease state has long term consequences in the heart or whether this represents a transition state where a greater BKCa current eventually compensates for elevated intracellular Ca^{2+} levels in the smooth muscle cells from the conduit arteries. It is conceivable that if the model was maintained for a longer time, the increase in BKCa currents would manifest in further altered coupling of intracellular Ca^{2+} or expose a limit to the ability to increase BKCa current, and start to accumulate intracellular Ca^{2+} with subsequently negative consequences. Because the exercise training reversed the altered functional coupling, this suggests that factors involved in BKCa coupling to intracellular Ca^{2+} might provide a novel therapeutic target. This includes BKCa channels, ryanodine sensitive sarcoplasmic reticulum channels, or other unidentified factors. This data contained in this review has provided some clarity regarding the regulation of KCa channels in coronary artery health and disease. There is most certainly a wealth of new insights that could be uncovered by study of the other major classes of K^+ channels and the plethora of isoforms.

Acknowledgments

This work was supported by National Institutes of Health grants HL062552 and P30 DK097512. Due to space limitations, we have not presented all articles in detail and we apologize for any outstanding work that we have not cited.

References

Aguilar-Bryan, L., Clement, J. P., Gonzalez, G., Kunjilwar, K., Babenko, A., & Bryan, J. (1998). Toward understanding the assembly and structure of K_{ATP} channels. *Physiological Reviews*, *78*, 227–245.

Albarwani, S., Al-Siyabi, S., Baomar, H., & Hassan, M. O. (2010). Exercise training attenuates ageing-induced BKCa channel downregulation in rat coronary arteries. *Experimental Physiology*, *95*(6), 746–755. https://doi.org/10.1113/expphysiol.2009.051250.

Benham, C. D., Bolton, T. B., Lang, R. J., & Takewaki, T. (1986). Calcium-activated potassium channels in single smooth muscle cells of rabbit jejunum and guinea pig mesenteric artery. *Journal of Physiology (London)*, *371*, 45–67.

Berne, R. M. (1980). The role of adenosine in the regulation of coronary blood flow. *Circulation Research*, *47*, 807–813.

Bowles, D. K., Laughlin, M. H., & Sturek, M. (1998). Exercise training increases K^+ channel contribution to regulation of coronary arterial tone. *Journal of Applied Physiology*, *84*(4), 1225–1233.

Brayden, J. E., & Nelson, M. T. (1992). Regulation of arterial tone by activation of calcium-dependent potassium channels. *Science*, *256*, 532–535.

Caesar, K., Akgoren, N., Mathiesen, C., & Lauritzen, M. (1999). Modification of activity-dependent increases in cerebellar blood flow by extracellular potassium in anaesthetized rats. *Journal of Physiology*, *520*(Pt 1), 281–292. https://doi.org/10.1111/j.1469-7793.1999.00281.x.

Daut, J., Maier-Rudolph, W., Von Beckerath, N., Mehrke, G., Günther, K., & Goedel-Meinen, L. (1990). Hypoxic dilation of coronary arteries is mediated by ATP- sensitive potassium channels. *Science*, *247*, 1341–1344.

Dick, G. M., & Tune, J. D. (2010). Role of potassium channels in coronary vasodilation. *Experimental Biology and Medicine*, *235*(1), 10–22.

Dogan, M. F., Yildiz, O., Arslan, S. O., & Ulusoy, K. G. (2019). Potassium channels in vascular smooth muscle: A pathophysiological and pharmacological perspective. *Fundamental and Clinical Pharmacology*, *33*(5), 504–523. https://doi.org/10.1111/fcp.12461.

Giangiacomo, K. M., Garcia, M. L., & McManus, O. B. (1992). Mechanism of iberiotoxin block of the large-conductance calcium-activated potassium channel from bovine aortic smooth muscle. *Biochemistry*, *31*(29), 6719–6727. https://doi.org/10.1021/bi00144a011.

Gordienko, D. V., Bolton, T. B., & Cannell, M. B. (1998). Variability in spontaneous subcellular calcium release in guinea-pig ileum smooth muscle cells. *Journal of Physiology*, *507*(Pt 3), 707–720. https://doi.org/10.1111/j.1469-7793.1998.707bs.x.

Hasdai, D., Nielsen, M. F., Rizza, R. A., Holmes, D. R., Jr., Richardson, D. M., Cohen, P., et al. (1999). Attenuated in vitro coronary arteriolar vasorelaxation to insulin-like growth factor I in experimental hypercholesterolemia. *Hypertension*, *34*(1), 89–95.

Heaps, C. L., Tharp, D., & Bowles, D. K. (2005). Hypercholesterolemia abolishes voltage-dependent k+ channel contribution to adenosine-mediated relaxation in porcine coronary arterioles. *American Journal of Physiology. Heart and Circulatory Physiology*, *288*, 568–576.

Inoue, R., Kitamura, K., & Kuriyama, H. (1985). Two ca-dependent K-channels classified by the application of tetraethylammonium distribute to smooth muscle membranes of the rabbit portal vein. *Pflügers Archiv - European Journal of Physiology*, *405*(3), 173–179. https://doi.org/10.1007/BF00582557.

Jackson, W. F. (2000). Ion channels and vascular tone. *Hypertension*, *35*(1), 173–178.

Jackson, W. F. (2005). Potassium channels in the peripheral microcirculation. *Microcirculation*, *12*(1), 113–127. https://doi.org/10.1080/10739680590896072.

Jaggar, J. H., Wellman, G. C., Heppner, T. J., Porter, V. A., Perez, G. J., Gollasch, M., et al. (1998). Ca^{2+} channels, ryanodine receptors and Ca^{2+}-activated K^+ channels: A functional unit for regulating arterial tone. *Acta Physiologica Scandinavica*, *164*, 577–587.

Kirber, M. T., Etter, E. F., Bellve, K. A., Lifshitz, L. M., Tuft, R. A., Fay, F. S., et al. (2001). Relationship of Ca2+ sparks to STOCs studied with 2D and 3D imaging in feline oesophageal smooth muscle cells. *Journal of Physiology, 531*(Pt 2), 315–327. https://doi.org/10.1111/j.1469-7793.2001.0315i.x.

Knot, H. J., & Nelson, M. T. (1995). Regulation of membrane potential and diameter by voltage- dependent K^+ channels in rabbit myogenic cerebral arteries. *American Journal of Physiology. Heart and Circulatory Physiology, 269*, H348–H355.

Knot, H. J., & Nelson, M. T. (1998). Regulation of arterial diameter and wall [Ca^{2+}] in cerebral arteries of rat by membrane potential and intravascular pressure. *Journal of Physiology (London), 508*, 199–209.

Knot, H. J., Standen, N. B., & Nelson, M. T. (1998). Ryanodine receptors regulate arterial diameter and wall [Ca^{2+}] in cerebral arteries of rat via Ca^{2+}-dependent K^+ channels. *Journal of Physiology (London), 508*, 211–221.

Ko, E. A., Han, J., Jung, I. D., & Park, W. S. (2008). Physiological roles of K+ channels in vascular smooth muscle cells. *Journal of Smooth Muscle Research, 44*(2), 65–81. https://doi.org/10.1540/jsmr.44.65.

Korovkina, V. P., & England, S. K. (2002). Molecular diversity of vascular potassium channel isoforms. *Clinical and Experimental Pharmacology and Physiology, 29*(4), 317–323. https://doi.org/10.1046/j.1440-1681.2002.03651.x.

Langton, P. D., Nelson, M. T., Huang, Y., & Standen, N. B. (1991). Block of calcium-activated potassium channels in mammalian arterial myocytes by tetra-ethylammonium ions. *American Journal of Physiology. Heart and Circulatory Physiology, 260*, H927–H934.

Martens, J. R., & Gelband, C. H. (1998). Ion channels in vascular smooth muscle: Alterations in essential hypertension. *Proceedings of the Society for Experimental Biology and Medicine, 218*, 192–203.

Merkus, D., Sorop, O., Houweling, B., Hoogteijling, B. A., & Duncker, D. J. (2006). KCa+ channels contribute to exercise-induced coronary vasodilation in swine. *AJP - Heart and Circulatory Physiology, 291*(5), H2090–H2097.

Miller, C., Moczydlowski, E., Latorre, R., & Phillips, M. (1985). Charybdotoxin, a protein inhibitor of single Ca2+—activated K+ channels from mammalian skeletal muscle. *Nature, 313*(6000), 316–318. https://doi.org/10.1038/313316a0.

Mironneau, J., Arnaudeau, S., Macrez-Lepretre, N., & Boittin, F. X. (1996). Ca^{2+} sparks and Ca^{2+} waves activate different Ca^{2+}-dependent ion channels in single myocytes from rat portal vein. *Cell Calcium, 20*(2), 153–160.

Mokelke, E. A., Dietz, N. J., Eckman, D. M., Nelson, M. T., & Sturek, M. (2005). Diabetic dyslipidemia and exercise affect coronary tone and differential regulation of conduit and microvessel K^+ current. *American Journal of Physiology. Heart and Circulatory Physiology, 288*(3), H1233–H1241. https://doi.org/10.1152/ajpheart.00732.2004.

Mokelke, E. A., Hu, Q., Song, M., Toro, L., Reddy, H. K., & Sturek, M. (2003). Altered functional coupling of coronary K^+ channels in diabetic dyslipidemic pigs is prevented by exercise. *Journal of Applied Physiology, 95*(3), 1179–1193. https://doi.org/10.1152/japplphysiol.00972.2002.

Nelson, M. T. (1993). Ca(2+)-activated potassium channels and ATP-sensitive potassium channels as modulators of vascular tone. *Trends in Cardiovascular Medicine, 3*(2), 54–60. https://doi.org/10.1016/1050-1738(93)90037-7.

Nelson, M. T., Cheng, H., Rubart, M., Santana, L. F., Bonev, A. D., Knot, H. J., et al. (1995). Relaxation of arterial smooth muscle by calcium sparks. *Science, 270*, 633–637.

Nelson, M. T., Patlak, J. B., Worley, J. F., & Standen, N. B. (1990). Calcium channels, potassium channels, and voltage dependence of arterial smooth muscle tone. *American Journal of Physiology. Cell Physiology, 259*, C3–C18. https://doi.org/10.1152/ajpcell.1990.259.1.C3.

Nelson, M. T., & Quayle, J. M. (1995). Physiological roles and properties of potassium channels in arterial smooth muscle. *American Journal of Physiology. Cell Physiology, 268*, C799–C822.

Park, W. S., Son, Y. K., Kim, N., Youm, J. B., Warda, M., Ko, J. H., et al. (2007). Direct modulation of ca(2+)-activated K(+) current by H-89 in rabbit coronary arterial smooth muscle cells. *Vascular Pharmacology, 46*(2), 105–113. https://doi.org/10.1016/j.vph.2006.08.413.

Sanchis-Gomar, F., Lavie, C. J., Marin, J., Perez-Quilis, C., Eijsvogels, T. M. H., O'Keefe, J. H., et al. (2022). Exercise effects on cardiovascular disease: From basic aspects to clinical evidence. *Cardiovascular Research, 118*(10), 2253–2266. https://doi.org/10.1093/cvr/cvab272.

Shi, L., Zhao, L., Zeng, F., Li, N., & Liu, X. (2012). Effect of exercise training volume on arterial contractility and BK(ca) channel activity in rat thoracic aorta smooth muscle cells. *European Journal of Applied Physiology, 112*(10), 3667–3678. https://doi.org/10.1007/s00421-012-2344-6.

Sieck, G. C., Kannan, M. S., & Prakash, Y. S. (1997). Heterogeneity in dynamic regulation of intracellular calcium in airway smooth muscle cells. *Canadian Journal of Physiology and Pharmacology, 75*(7), 878–888. Retrieved from https://www.ncbi.nlm.nih.gov/pubmed/9315357.

Snyders, D. J. (1999). Structure and function of cardiac potassium channels. *Cardiovascular Research, 42*(2), 377–390.

Somlyo, A. P., & Somlyo, A. V. (1994). Signal transduction and regulation in smooth muscle. *Nature, 372*, 231–236.

Sorensen, C. M., Braunstein, T. H., Holstein-Rathlou, N. H., & Salomonsson, M. (2012). Role of vascular potassium channels in the regulation of renal hemodynamics. *American Journal of Physiology. Renal Physiology, 302*(5), F505–F518. https://doi.org/10.1152/ajprenal.00052.2011.

Standen, N. B., Quayle, J. M., Davies, N. W., Brayden, J. E., Huang, Y., & Nelson, M. T. (1989). Hyperpolarizing vasodilators activate ATP-sensitive K^+ channels in arterial smooth muscle. *Science, 245*, 177–180.

Sturek, M. (2011). Ca^{2+} regulatory mechanisms of exercise protection against coronary artery disease in metabolic syndrome and diabetes. *Journal of Applied Physiology, 111*(2), 573–586. https://doi.org/10.1152/japplphysiol.00373.2011.

Tanaka, Y., Koike, K., & Toro, L. (2004). MaxiK channel roles in blood vessel relaxations induced by endothelium-derived relaxing factors and their molecular mechanisms. *Journal of Smooth Muscle Research, 40*(4–5), 125–153. https://doi.org/10.1540/jsmr.40.125.

Toro, L., Vaca, L., & Stefani, E. (1991). Calcium-activated potassium channels from coronary smooth muscle reconstituted in lipid bilayers. *American Journal of Physiology. Heart and Circulatory Physiology, 260*, H1779–H1789.

von Beckerath, N., Cyrys, S., Dischner, A., & Daut, J. (1991). Hypoxic vasodilatation in isolated, perfused guinea-pig heart: An analysis of the underlying mechanisms. *Journal of Physiology, 442*, 297–319. https://doi.org/10.1113/jphysiol.1991.sp018794.

Waldron, G. J., & Cole, W. C. (1999). Activation of vascular smooth muscle K^+ channels by endothelium-derived relaxing factors. *Clinical and Experimental Pharmacology & Physiology, 26*(2), 180–184.

CHAPTER SEVEN

K^+ channels in the coronary microvasculature of the ischemic heart

Sharanee P. Sytha[a,†], Trevor S. Self[a,†], and Cristine L. Heaps[a,b,*]

[a]Department of Physiology and Pharmacology, School of Veterinary Medicine and Biomedical Sciences, Texas A&M University, College Station, TX, United States

[b]Michael E. DeBakey Institute for Comparative Cardiovascular Science and Biomedical Devices, School of Veterinary Medicine and Biomedical Sciences, Texas A&M University, College Station, TX, United States

*Corresponding author: e-mail address: cheaps@tamu.edu

Contents

Abstract

Ischemic heart disease is the leading cause of death and a major public health and economic burden worldwide with expectations of predicted growth in the foreseeable future. It is now recognized clinically that flow-limiting stenosis of the large coronary conduit arteries as well as microvascular dysfunction in the absence of severe stenosis can each contribute to the etiology of ischemic heart disease. The primary site of coronary vascular resistance, and control of subsequent coronary blood flow, is found in the coronary microvasculature, where small changes in radius can have profound impacts on myocardial perfusion. Basal active tone and responses to vasodilators and vasoconstrictors are paramount in the regulation of coronary blood flow and adaptations in signaling associated with ion channels are a major factor in determining alterations in vascular resistance and thereby myocardial blood flow. K^+ channels are of particular importance as contributors to all aspects of the regulation of arteriole resistance and control of perfusion into the myocardium because these channels dictate

† These authors contributed equally.

Current Topics in Membranes, Volume 90
ISSN 1063-5823
https://doi.org/10.1016/bs.ctm.2022.09.004

membrane potential, the resultant activity of voltage-gated calcium channels, and thereby, the contractile state of smooth muscle. Evidence also suggests that K^+ channels play a significant role in adaptations with cardiovascular disease states. In this review, we highlight our research examining the role of K^+ channels in ischemic heart disease and adaptations with exercise training as treatment, as well as how our findings have contributed to this area of study.

1. Introduction

Cardiovascular disease is the leading cause of death worldwide with ischemic heart disease persisting as the most prevalent form (GBD 2015 Disease and Injury Incidence and Prevalence Collaborators, 2016). In physiological settings, the coronary circulation supplies the myocardium with blood based upon the metabolic demands of the heart (Duncker, Koller, Merkus, & Canty, 2015). Because oxygen extraction is high under resting conditions, increases in metabolic activity, such as during exercise, are met primarily by vasodilation of the coronary vasculature to increase coronary blood flow proportionally to the demands of the myocardium. Pathophysiological conditions, such as ischemia, where there is inadequate blood flow to the myocardium, can be brought about by the build-up and rupture of atherosclerotic plaques, severe occlusion of an epicardial conduit artery, or microvascular dysfunction, resulting in a mismatch between metabolic needs and perfusion of the myocardium. Oftentimes, patients that experience angina yet have no angiographic evidence for significant obstructive coronary lesions or macrovascular (conduit artery) spasm, display functional and structural impairments of the coronary microcirculation which lead to impaired vasodilation and thereby, myocardial ischemia (Lee et al., 2022). Alternatively, coronary microvascular spasm can also occur, resulting in significant reductions in myocardial blood flow. Mechanisms that contribute to coronary microvascular dysfunction may include an imbalance between vasodilators and vasoconstrictors released by the endothelium or enhanced tone of the vascular smooth muscle that may occur independently of endothelium-derived vasodilators (Vancheri, Longo, Vancheri, & Henein, 2020).

Improvements in perfusion of the ischemic myocardium can be brought about by increases in collateral development as well as enhanced vasodilation responses of the innate microvasculature in the ischemic regions of the heart. There is substantial evidence that exercise training is a valuable, non-pharmacological therapeutic option for cardiovascular disease (Duncker & Bache, 2008; Griffin, Laughlin, & Parker, 1999; Griffin, Woodman, Price,

Laughlin, & Parker, 2001). A significant body of literature has supported a role for physical activity in patients with coronary artery disease to induce improvements in coronary endothelial function and endothelium-dependent relaxation in the diseased coronary vasculature. Early work by Hambrecht and colleagues (Hambrecht et al., 2000) revealed that after completion of a 4-week cycle ergometry exercise regimen, atherosclerotic coronary disease patients displayed improved endothelium-dependent vasodilation in conduit coronary arteries and arterioles and enhanced coronary flow reserve. Considerable evidence over the years has confirmed these data and repeatedly shown enhanced vasodilation capacity in diseased patients and animal models after exercise training (Conraads et al., 2015; Koller et al., 2022). In contrast, direct evidence of increased collateralization with exercise training is limited in human patients with coronary artery disease, although animal models support a role for collateral expansion with exercise (Bloor, White, & Sanders, 1984; Cohen, Yipintsoi, & Scheuer, 1982; Roth et al., 1990). Numerous studies have reported no increase in collateralization after exercise training of patients (Merkus, Muller-Delp, & Heaps, 2021). Only a small number of studies have provided direct angiographic evidence that exercise training potentiated coronary collateralization along with improved perfusion by collateral vessels and improvements in myocardial contractility in diseased patients (Belardinelli, Belardinelli, & Shryock, 2001; Belardinelli, Georgiou, Ginzton, Cianci, & Purcaro, 1998). However, because assessment of collateralization by angiography has significant limitations in detection of vessels smaller than ~200 μm, the assessment of collateral development is insufficient in humans. Thus, despite exercise training-induced improvements in contractility of myocardial regions compromised by ischemia, the relative contributions of collateralization and enhanced reactivity of the microvasculature supplying that region remain unclear.

Coronary vascular tone is highly regulated in order to match oxygen delivery with metabolic demand. Vascular smooth muscle cells comprise the majority of the vascular wall and dictate blood flow by changes in contractile state and thus lumen diameter. Indeed, consistent with Pouseille's Law, small changes in vascular radius markedly impact blood flow into the dependent region. Vascular smooth muscle cells express a variety of ion channels that impact cellular membrane potential, with K^+ channels having the largest impact (Tykocki, Boerman, & Jackson, 2017). There is a vast number of K^+ channel classes, each with numerous isoforms, including voltage-gated K^+ (Kv) channels, calcium-activated K^+ (KCa) channels, adenosine triphosphate-sensitive K^+ (KATP) channels, and inward-rectifying K^+ (KIR) channels (Fig. 1). Numerous studies have identified members of

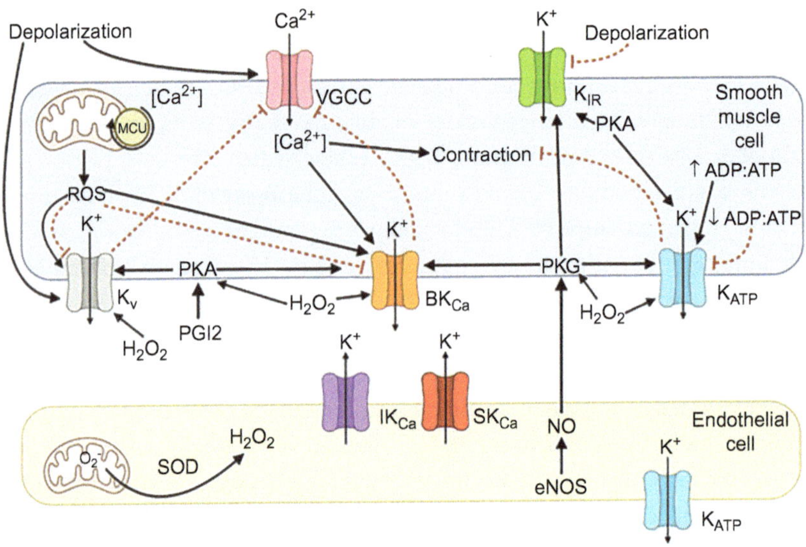

Fig. 1 Potassium channel pathways in the vascular smooth muscle and endothelium. Schematic depicts depolarization inducing activation of voltage-gated calcium channels (VGCC) and voltage-gated potassium (Kv) channels as denoted by arrows. Activation of VGCC allows for calcium movement into the cell, stimulating contraction. Large-conductance calcium-activated potassium (BKCa) channels open in response to increased intracellular calcium, allowing for K^+ efflux to counteract the depolarization and contractile response of calcium influx. Kv channel activation via depolarization leads to K^+ efflux. Together, Kv and BKCa channels allow K^+ efflux to hyperpolarize the cell membrane, combating VGCC stimulation, inducing relaxation of the contractile machinery. Intracellular calcium is also transported into the mitochondria via mitochondrial calcium uniporters as depicted in the smooth muscle cell. The reactive oxygen species (ROS) produced by the mitochondria can both stimulate and inhibit the activity of the Kv and BKCa channels. Within the smooth muscle cell are PKA and PKG proteins, which can be activated by PGI2 and NO respectively, that have been found to stimulate numerous potassium channels. In addition to PKA and PKG, the KATP channel is dependent on an increased ADP:ATP ratio for stimulation, while decreased ADP:ATP leads to inhibition. The stimulated KATP channel allows for outward potassium transport, resulting in inhibition of contraction. On the endothelial cell, there are the intermediate and small conductance calcium-activated potassium channels as well as KATP channels. Depolarization inhibits inward-rectifying potassium channel (KIR) as indicated with the dashed blunt arrow. Endothelial cell mitochondria can produce ROS (e.g., superoxide) which is then converted to hydrogen peroxide via superoxide dismutase (SOD). Hydrogen peroxide can activate various potassium channels. Under ischemic conditions, it has been proposed that numerous potassium channels including Kv, BKCa, and KATP, may act as redox or oxygen sensors, involved in regulation of coronary blood flow. See text for further details and references. Created with Biorender.com.

each class within the coronary smooth muscle, most notably, Kv, KATP, KIR, and large-conductance, calcium-dependent KCa (BKCa) channels (Tykocki et al., 2017).

Smooth muscle resting membrane potential is largely controlled by K^+ leak channels on the membrane surface with Kv and BKCa channels also playing a role (Gutterman, Miura, & Liu, 2005). Depolarization of the smooth muscle plasma membrane leads to activation of Kv channels and subsequent hyperpolarization. This reduces the activity of voltage-gated Ca^{2+} channels, diminishing intracellular Ca^{2+} concentrations, resulting in smooth muscle relaxation and vasodilation (Kim, Appel, Vetterkind, Gangopadhyay, & Morgan, 2008). In parallel, increased intracellular Ca^{2+} concentrations, resulting from membrane depolarization, activate KCa channels yielding membrane hyperpolarization, diminishing intracellular Ca^{2+} concentrations via closure of voltage-gated Ca^{2+} channels, and subsequent vasodilation (Berkefeld, Fakler, & Schulte, 2010; Hill, Yang, Ella, Davis, & Braun, 2010). Due to the high membrane resistance of vascular smooth muscle cells, small changes in K^+ channel activity can have significant impact on arteriolar tone (Nelson, Patlak, Worley, & Standen, 1990). In addition, numerous K^+ channels including Kv channels of the Kv1.5 and Kv2.1 subtype (Dwenger, Ohanyan, Navedo, & Nystoriak, 2018; Nishijima et al., 2017; Ohanyan et al., 2015; Rogers et al., 2006; Rogers, Chilian, Bratz, Bryan, & Dick, 2007; Shimizu, Yokoshiki, Sperelakis, & Paul, 2000; Thorne, Conforti, & Paul, 2002), BKCa (Gebremedhin et al., 1994; López-Barneo, del Toro, Levitsky, Chiara, & Ortega-Sáenz, 2004; Nelson & Quayle, 1995), and KATP (Dart & Standen, 1995; Kamekura et al., 1999; Lee, Kim, & Kang, 1998) channels have been proposed as redox or O_2 sensors and involved in coronary vasodilation associated with ischemic conditions. With each of these subfamilies possessing multiple isoforms, identifying which K^+ channels are involved in ischemia/hypoxia-induced vasodilation is complex. Further, ion channel expression and regulation are different in macro- vs. microvasculature as shown by greater frequency of spontaneous transient outward currents (BKCa) (Mokelke, Dietz, Eckman, Nelson, & Sturek, 2005) and voltage-gated Ca^{2+} channels (Bowles, Hu, Laughlin, & Sturek, 1998) in microvascular compared to macrovascular smooth muscle. These differences point to the specificity of coronary microvascular regulation. Thus, exploration of the contribution of specific K^+ channel isoforms is actively being pursued in microvascular smooth muscle and much insight remains to be gained from future study of the role of specific channel isoforms in coronary blood flow under control and compromised conditions as well as adaptations associated with exercise and other therapies.

Another family of potassium channels that have an effect on vascular tone are the inward-rectifying potassium (KIR) channels, expressed in both smooth muscle and endothelial cells (Hibino et al., 2010; Quayle, Nelson, & Standen, 1997). Under physiological conditions, the inward-rectifying potassium channel (KIR2.1) regulates vasodilation via nitric oxide production (Ahn et al., 2017; Fancher et al., 2018). Nelson and colleagues (Sonkusare, Dalsgaard, Bonev, & Nelson, 2016) determined that endothelial cell KIR2 channels may amplify the vasodilation signals of resistance arteries produced by SKCa and IKCa channels. Others have shown a small role for endothelial cell KIR channels in dilation of human coronary arterioles in response to slight elevations in extracellular potassium levels (Miura, Toyama, Pratt, & Gutterman, 2011), such as those observed during exercise (Wilson, Kapoor, & Krishna, 1994). The role of these channels in patients and animal models of ischemic heart disease have not been explored in detail.

The focus of this review is to highlight our research examining the role of microvascular smooth muscle K^+ channels in ischemic heart disease and how our research has contributed to the field of study. We will focus primarily on adaptations in the coronary microcirculation distal to epicardial stenosis/occlusion as a key mechanism regulating coronary ischemia.

2. Porcine model of ischemic heart disease

Our laboratory has had a long-standing interest in the mechanisms by which exercise training drives adaptations in the microcirculation of the ischemic heart that increase blood flow into compromised myocardium and thereby, improve the function (e.g., contractility) of the heart. Indeed, the regulation of blood flow through the coronary microvasculature determines function of not only the heart but the entire body, since inadequate myocardial blood flow negatively impacts the ability of the heart to pump effectively, limiting delivery of blood to the periphery. Over the years, we have examined exercise training-induced adaptations of the coronary microcirculation using porcine models of both hypercholesterolemia and chronic coronary artery occlusion.

We use a porcine model of chronic coronary artery occlusion and exercise training to gain a better understanding of the fundamental cellular and molecular mechanisms that underlie exercise-induced cardioprotection.

Our goal is to provide insight into new therapeutic targets for increasing blood flow into compromised myocardium. For much of the work we will discuss in this review, we have used sexually-mature Yucatan miniature swine (6–7 months of age) surgically instrumented with an ameroid constrictor around the proximal left circumflex coronary artery. Animals are placed under general anesthesia (2–3% isoflurane and supplemental O_2) during aseptic surgery and a left lateral thoracotomy is performed in the fourth intercostal space. The underlying pericardium is opened to expose the proximal left circumflex artery, which is then dissected free of surrounding tissue and an ameroid constrictor (2.5- to 3.5-mm inner diameter) is placed around the artery (Fig. 2A). The diameter of the occluder is selected upon visual inspection of the artery to initially provide a secure but non-constrictive fit. The pericardium is then closed and the thoracotomy is repaired in tissue layers. Pigs are closely monitored during surgery and surgical recovery, receiving pain medication as needed. Ameroid closure and total occlusion

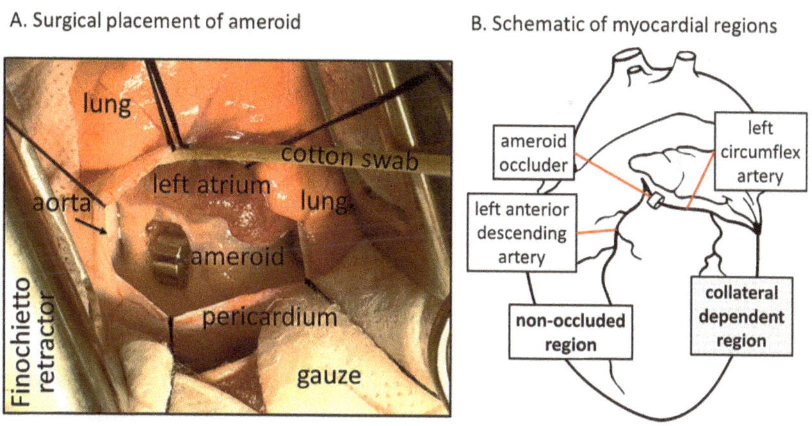

Fig. 2 (A) During aseptic surgery, a left lateral thoracotomy is performed in the fourth intercostal space. Finochietto retractor (pediatric) is used to maintain a clear surgical field. The underlying pericardium is opened to expose the myocardial region containing the proximal LCX artery. The LCX artery is dissected free of surrounding tissue and an ameroid constrictor (stainless steel device) is placed around the artery. Pericardium is held in the open position during placement of the ameroid using silk suture (2-0). Gauze is used beneath the silk suture to protect underlying tissues. (B) 3 weeks post-operatively, complete occlusion of the LCX artery is achieved, generating a downstream, collateral-dependent myocardium. This tissue is now reliant on expansion of the collateral circulation supported by flow from both the nonoccluded left anterior descending (LAD) and right coronary (RCA) arteries. The myocardium supplied by the LAD is defined as the nonoccluded control region.

ensues approximately 3 weeks after ameroid placement (White, Carroll, Magnet, & Bloor, 1992). Pigs undergo postoperative recovery for 8 weeks before the sedentary or exercise training experimental protocol is initiated. Previous studies have demonstrated that between 3 and 7 weeks following ameroid placement, the development of collateral vessels in response to occlusion restores cardiac function under resting conditions and further collateral development is not observed (Roth et al., 1987). Thus, collateral vessel development in response to exercise can be measured independent of surgically-induced collateral growth. The schematic in Fig. 2B shows the placement of the ameroid occluder around the proximal left circumflex artery and the consequent collateral-dependent region produced after complete occlusion, as well as the nonoccluded region served by the left anterior descending artery.

This porcine model is widely recognized as highly clinically relevant because it replicates many human adaptations to both ischemic heart disease and exercise training (Cohen, 1985; Douglas, 1972; O'Konski, White, Longhurst, Roth, & Bloor, 1986; Roth et al., 1987, 1990; Schaper, 1971a; White et al., 1992; White & Bloor, 1981, 1992; White, Roth, & Bloor, 1986). The cardiovascular system and in particular the coronary circulation of the pig has many anatomic and physiologic attributes that mimic those of humans (Douglas, 1972; Schaper, 1971a; White & Bloor, 1981). Furthermore, the pig is an excellent model of progressive coronary artery occlusion because development of collateral vessels in the pig heart is sufficient to provide normal blood flow to the collateral-dependent myocardium at rest; however, blood flow during stress (e.g., exercise) remains compromised and unable to support regional myocardial function (O'Konski et al., 1986; Roth et al., 1987, 1990; White et al., 1986, 1992). Similarly, the human heart demonstrates few innate collateral vessels and growth of new vessels typically occurs as an extensive network of functionally significant collaterals in the endocardial and mid-myocardial layers (Cohen, 1985; White & Bloor, 1992; White et al., 1992). Furthermore, coronary artery disease patient populations often lack clinical symptoms at rest, but exhibit significant signs of persistent regional myocardial ischemia and contractile dysfunction (angina, ECG abnormalities) during periods of increased cardiac workload (Kolibash et al., 1982; Niebauer et al., 1995; Schuler et al., 1992) similar to that observed in pigs (O'Konski et al., 1986; Roth et al., 1987; White et al., 1992). Finally, histological analyses demonstrate that this porcine model displays an average infarct size of only ~1% of the left ventricular myocardium (Roth et al., 1987, 1990), thus, this model is an effective model of ischemic heart disease.

In contrast to humans and pigs, the canine heart possesses an abundance of innate collateral vessels, which can be recruited rapidly under conditions of ischemia (Cohen, 1985; Schaper, 1971b). Furthermore, collateralization of the canine heart generally occurs as an epicardial collateral network in contrast to the endocardial distribution in humans and pigs. The combination of vessel recruitment and the growth of new collaterals are sufficient to restore adequate blood flow to the compromised myocardium during rest and often under conditions of increased myocardial oxygen demand (Bache & Schwartz, 1983; Cohen & Yipintsoi, 1981; Eckstein, 1957; Lambert, Hess, & Bache, 1977; Longhurst, Motohara, Atkins, & Ordway, 1985). Thus, study of vascular adaptations to ischemia in the canine model may not be highly translatable to humans because the dog does not experience the level of ischemia found with obstructive coronary artery disease observed in human patients. Alternatively, the rapid collateral vessel growth observed in the dog may serve as an advantage in elucidating mechanisms of collateral development that can be exploited therapeutically to improve collateral development in humans. Rodents (rats and mice) are clearly inadequate for coronary microvascular responses to occlusion and exercise because the high resting heart rate indicates very high metabolic rate and the minimal changes in heart rate during exercise are vastly different than that observed in humans (Duncker & Bache, 2008).

In addition to similarities in response to occlusive artery disease, the porcine model also mimics responses to exercise training in that pigs are able to increase their maximal oxygen consumption with persistent exercise training and the maximal capacity of the coronary circulation is similar to that of humans (White et al., 1986). Furthermore, analogous to that observed in humans, swine demonstrate redistribution of blood flow from visceral tissue to skeletal muscle during exercise and display similar maximal oxygen consumption during exercise (Armstrong, Delp, Goljan, & Laughlin, 1987). Furthermore, despite perfusion deficits during increased cardiac workload (Roth et al., 1987), persistent exercise training has been shown to improve the perfusion deficit and contractile dysfunction of the myocardium at risk (Roth et al., 1990).

To confirm previous reports (O'Konski et al., 1986; Roth et al., 1987, 1990; White et al., 1986; White & Bloor, 1992), we performed dobutamine stress echocardiography to examine regional myocardial wall function in our chronically occluded swine using peak systolic myocardial velocity as measure of function (Fig. 3). Our data indicate that under resting conditions, myocardial function is not impaired in the collateral-dependent region compared with the nonoccluded region. In contrast, at high dobutamine doses

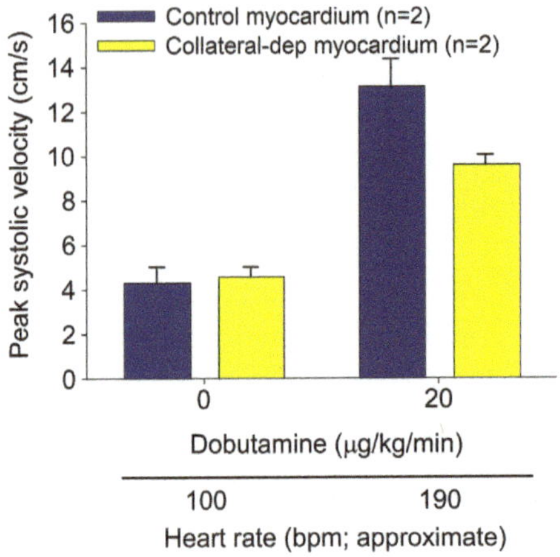

Fig. 3 Regional myocardial wall function as determined by dobutamine stress echocardiography using peak systolic myocardial velocity as measure of function. Wall function was determined in both the myocardial region supplied by the left anterior descending artery (nonoccluded) and the collateral-dependent region previously supplied by the left circumflex coronary artery. At rest (0 μg/kg/min dobutamine), peak velocity is similar between the nonoccluded and collateral-dependent regions, whereas during increased demand (20 μg/kg/min dobutamine), peak velocity is markedly reduced in the collateral-dependent compared with the nonoccluded region.

(20 g/kg/min), when heart rates of ~190 bpm are attained, function of the collateral-dependent region is markedly diminished compared with the nonoccluded region. This finding is of great significance because during each bout of treadmill exercise, our exercise-trained animals attain heart rates of ~190–200 bpm (~65–70% of maximal heart rate), suggesting that during exercise, our pigs likely experience ischemic episodes that may contribute to adaptations of the collateral-dependent region. Completion of such studies under cardiac stress are important since patients with advanced coronary artery disease often lack clinical symptoms at rest, but exhibit significant signs of myocardial ischemia (angina, ECG abnormalities) during periods of increased cardiac workload. Taken together, these data indicate that the porcine model of progressive, chronic occlusion is a representative and valid model of stress-induced ischemic heart disease that closely mimics human ischemic heart disease.

Serial coronary angiographic images (Fig. 4) taken 21-weeks following placement of the ameroid constrictor confirm that blood flow through the left circumflex artery at the site of occlusion was completely abolished by the ameroid constrictor. Relatively large, tortuous collateral vessels supplying

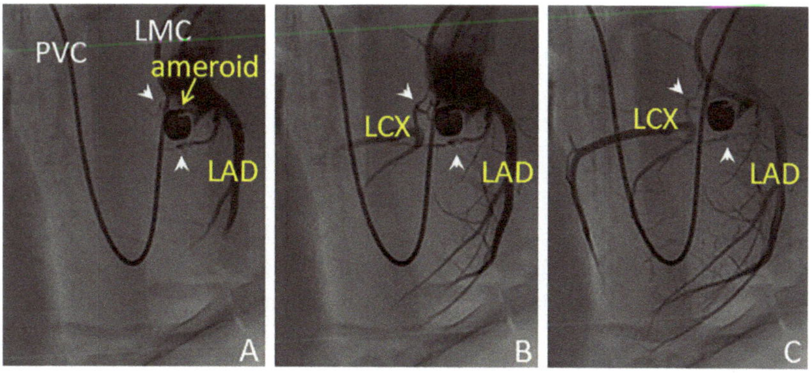

Fig. 4 Sequential angiograms of the porcine model of progressive chronic coronary artery occlusion and collateral development. 22 weeks after surgical placement of an ameroid occluder around the proximal left circumflex coronary artery, the pig was placed under general anesthesia and hemodynamics were monitored throughout the procedure. Arterial access was obtained by surgical cutdown of the carotid artery. The left main coronary artery was catheterized with a 6F guiding catheter (Vista Brite Tip; Cordis) introduced over a 0.035-in. guidewire. Selective coronary angiography was performed with nonionic contrast agent (Oxilan; 350mgL/mL). (A–C): regionally enhanced serial images emphasizing the cardiac silhouette to provide additional detail of coronary vasculature. White arrowheads identify collateral vessels supplying the left circumflex artery (LCX) distal to occlusion. The left anterior descending (LAD) artery demonstrates unobstructed flow. LMC, left main coronary artery catheter; PVC, pulmonary vein catheter. *Adapted from Heaps, C. L., & Parker, J. L. (2011). Effects of exercise training on coronary collateralization and control of collateral resistance.* Journal of Applied Physiology, 111*(2), 587–598.*

the left circumflex artery distal to occlusion are clearly visible in both panels of the angiogram. These collaterals appear to originate from the proximal left anterior descending and the left circumflex artery (proximal to occluder) and "bypass" the occlusion. Smaller intramyocardial collaterals are likely also present (White et al., 1992) but are difficult to visualize by angiography (Heaps & Parker, 2011). Importantly, because complete ameroid-induced left circumflex artery occlusion is evident, the myocardial region formerly perfused by the native left circumflex artery is, by definition, now fully dependent upon perfusion via collateral vessels such as those observed in Fig. 4.

3. Role of K^+ channels in basal active tone of arterioles from ischemic myocardium

Our early work evaluating the coronary microcirculation in ischemic heart disease explored the effect of exercise training on K^+ channels as mediators of basal active tone in the microvasculature distal to chronic occlusion. These studies revealed that 4-aminopyridine-sensitive Kv channels contribute markedly to resting tone in arterioles of nonoccluded and collateral-dependent myocardial regions in both sedentary and exercise-trained animals. Intriguingly, the contractile response to Kv channel blockade was statistically greater in arterioles from the collateral-dependent region of exercise-trained pigs, indicating that Kv channels contribute to the regulation of basal tone to a greater extent in arterioles distal to occlusion after exercise training (Heaps, Mattox, Kelly, Meininger, & Parker, 2006). Kv channels that have been shown to be 4-aminopyridine-sensitive include primarily the Kv1-Kv4 subfamilies (Tykocki et al., 2017), suggesting that some members of these channel subtypes contribute significantly to basal active tone in the coronary circulation with marked induction in the collateral-dependent region after exercise training. On the other hand, BKCa channel contribution to basal tone was negligible in all vessel treatment groups (Heaps et al., 2006), a finding that was also demonstrated in large conduit arteries from these animals (Deer & Heaps, 2013). In a previous study, Bowles, Laughlin, and Sturek (1998) demonstrated an increased role for both 4-AP-sensitive Kv channels and BKCa channels in the maintenance of basal tone after exercise training in porcine epicardial conduit artery segments, although the role of 4-AP-sensitive Kv channels was markedly greater than that of BKCa channels.

Paradoxically, in addition to an increased contribution of K^+ channels to basal tone, we also observed enhanced Ca^{2+}-dependent basal active tone and increased nitric oxide contributions to resting tone (Heaps et al., 2006). Taken together, these data suggest that exercise training drives a parallel upregulation of seemingly contradictory signaling mechanisms that may ultimately function to optimize the control of coronary basal tone. Adaptations in both vasodilation and vasoconstriction pathways may more precisely maintain appropriate blood flow to meet the metabolic demands of the at-risk myocardium, as well as contribute to the maintenance of coronary flow reserve. The stimulus for these adaptations may be most marked in the collateral-dependent region because of coupling of intermittent ischemia with mechanical forces on the vascular wall during each bout of exercise. Indeed, exercise-induced myocardial ischemia persists in this animal model despite improvements in coronary blood flow and myocardial contractile function after chronic exercise training (Roth et al., 1990).

4. Kv channels in arteriolar smooth muscle cells of ischemic myocardium

In subsequent studies, we evaluated the effects of chronic coronary occlusion and exercise training on whole-cell K^+ currents in freshly isolated coronary arteriolar smooth muscle cells from nonoccluded and collateral-dependent myocardial regions (Xie, Parker & Heaps, 2013). We revealed that K^+ channel currents were significantly decreased in arteriolar smooth muscle cells from collateral-dependent compared with nonoccluded regions in both sedentary and exercise-trained pigs. Interestingly, after exercise training, cells from arterioles in the collateral-dependent region displayed significantly reduced whole cell K^+ channel currents compared with those in sedentary pigs and a strong tendency ($p = 0.07$) for reduced K^+ channel currents in the nonoccluded region also was apparent after exercise. Thus, both chronic occlusion and exercise training resulted in a reduction in whole cell K^+ channel currents, which was further diminished in the presence of occlusion plus exercise training. In contrast with these findings, smooth muscle cells isolated from conduit arteries demonstrated that neither exercise training nor chronic coronary occlusion altered whole cell K^+ currents. Previous studies have shown unique exercise training-induced adaptations of K^+ channel currents in coronary arteries compared with coronary arterioles of diabetic, dyslipidemic pigs (Mokelke et al., 2005; Mokelke et al., 2003). These studies have reported a reversal of increased K^+ channel current in conduit smooth muscle cells after exercise training (Mokelke et al., 2003) and an increase in spontaneous transient outward currents in smooth muscle of intact arteriole segments after exercise (Mokelke et al., 2005). Similar to our findings in control and collateral-dependent conduit arteries (Deer & Heaps, 2013), other investigators using healthy control animals, have shown no effect of exercise training on whole-cell K^+ channel currents in smooth muscle cells isolated from conduit coronary arteries, although both BKCa and 4-AP-sensitive Kv channels contributed more to basal tone in exercise-trained compared with sedentary animals (Bowles, Laughlin, et al., 1998).

Vascular smooth muscle cells in the coronary circulation express a wide range of Kv channels, with the Kv1, Kv2, and Kv7 families regarded as the most significant (Jackson, 2018). The wide variety of Kv channel isoforms likely makes them suitable end effectors in response to many different stimuli. Variations in targets across the different Kv channel isoforms may be due to differential subcellular localization or different binding sites located on specific channels. It has been recently proposed that the Kv7 family is involved in ischemia-induced vasodilation by means of H_2S and adenosine stimulation in a porcine model (Hedegaard et al., 2014). However, others have reported that Kv7 channels are not involved in metabolic or ischemic regulation of coronary vasodilation, but instead contribute to paracrine regulation via endothelium dependent mechanisms in swine (Goodwill et al., 2016). Interestingly, exercise training reversed a reduction in Kv1.2 and Kv1.5 channel expression in spontaneously hypertensive rats (Li, Lu, & Shi, 2014). Nishijima previously reported a loss of 4AP-sensitive Kv channel contribution to dilation in adipose tissue arterioles from coronary artery disease patients compared with those from non-diseased patients (Nishijima et al., 2017). The observed reduction in Kv channel contribution to dilation was attributed to a loss of Kv1.5 isoform in this study and a shift to BKCa as the predominant end effector in diseased patients (Nishijima et al., 2017). The same group also reported a reduced role for 4AP-sensitive Kv channels to dilation, and specifically the Kv1.5 isoform, in coronary arterioles isolated from diseased patients compared with those from patients not suffering from coronary artery disease (Nishijima et al., 2018).

5. BKCa channels in arteriolar smooth muscle cells of ischemic myocardium

It has been repeatedly shown that BKCa channels are abundantly expressed in vascular smooth muscle cell membranes and that they act as a negative feedback to increased intracellular calcium concentrations to reduce arterial tone (Berkefeld et al., 2010; Hill et al., 2010; Rusch, 2009; Toro et al., 2014). Under physiological conditions, receptor-mediated vasoconstriction is reduced through increases in BKCa channel activity (Albarwani, Al-Siyabi, Baomar, & Hassan, 2010; Bowles, Laughlin, et al., 1998; Laughlin, Bowles, & Duncker, 2012). Closely related to our work, previous studies in coronary arterioles of coronary artery disease patients demonstrated that both flow-mediated (Miura et al., 2001) and H_2O_2-induced (Liu, Bubolz, Mendoza, Zhang, & Gutterman, 2011; Zhang et al., 2012) dilation are mediated in part by BKCa channels. We were interested in exploring if the contribution of BKCa channels to endothelium-dependent dilation was an adaptation of the diseased state since this signaling pathway was not explored previously in non-diseased arterioles and whether the exercise training-enhanced dilation observed previously by our laboratory (Xie et al., 2013) was attributable to BKCa channel activation.

We examined a role for BKCa channels in endothelium-dependent dilation of the coronary microvasculature as well as that of large conduit coronary arteries. In cannulated coronary arterioles, endothelium-dependent dilation was impaired in the collateral-dependent region of ameroid-occluded swine (Xie, et al., 2013). Exercise training restored the impaired dilation, which we revealed was at least partially dependent on a greater contribution of BKCa channels after exercise. Previous literature also shows that increases in BKCa channel activity is brought about by exercise training in a porcine model of heart failure (Olver et al., 2018). We also evaluated whole-cell iberiotoxin-sensitive K^+ channel current in freshly isolated coronary arteriolar smooth muscle cells from nonoccluded and collateral-dependent myocardial regions. These data revealed no differences in iberiotoxin-sensitive currents. Using immunoblot analysis, we have also discovered that protein levels for BKCa channel α-subunit did not differ in arterioles from nonoccluded and collateral-dependent arterioles of sedentary and exercise-trained swine (data not published). The lack of differences in BKCa channel currents and BKCa channel protein levels across treatment groups suggest that the increased contribution of BKCa channels after exercise in collateral-dependent arterioles may reflect better coupling of BKCa channels with second-messenger signaling pathways (e.g., Ca^{2+}) that utilize BKCa channel as an end effector. We presume similar adaptations with Kv channel subtypes, which will be an area of future study in our laboratory.

We also demonstrated in freshly isolated smooth muscle cells of conduit coronary arteries there were no differences in iberiotoxin-sensitive K^+ currents (Deer & Heaps, 2013). Interestingly, we demonstrate for the first time in Fig. 5 the comparison of K^+ channel currents in smooth muscle cells isolated from conduits and arterioles. These comparisons reveal that total whole-cell K^+ channel currents in smooth muscle cells isolated from conduit arteries are relatively smaller than those from arterioles in both nonoccluded (Fig. 5A) and collateral-dependent (Fig. 5B) vessels. In contrast, iberiotoxin-sensitive K^+ currents are relatively greater in conduit than arteriole smooth muscle cells of both nonoccluded (Fig. 5C) and collateral-dependent (Fig. 5D) vessels. Taken together, these data suggest that Kv channels may play a more prominent role in coronary arterioles while BKCa channels may play a more prominent role in larger arteries. It should be noted that in the experiments performed for Fig. 5, pipettes were filled with

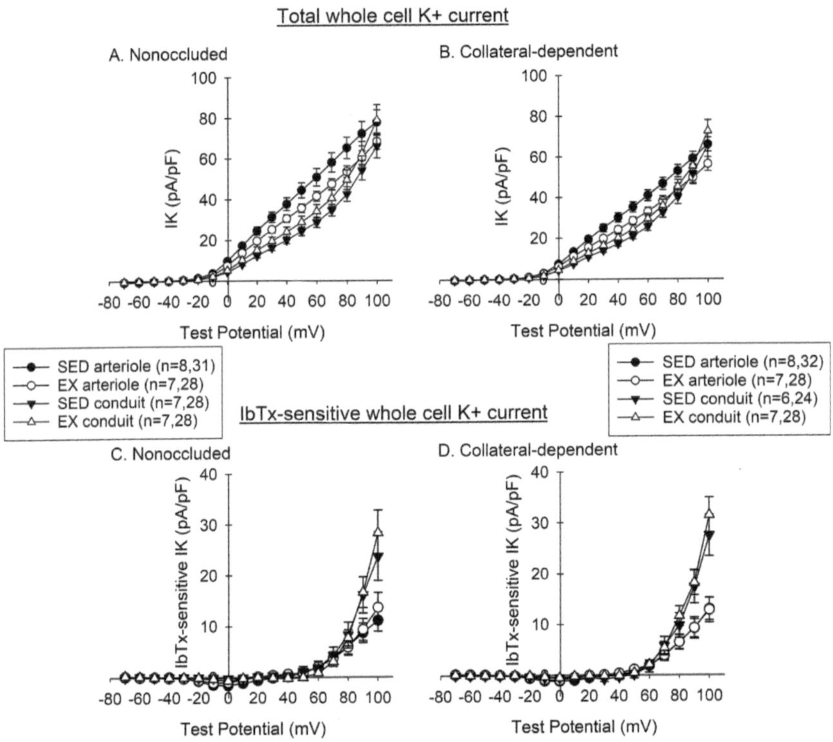

Fig. 5 Effect of chronic occlusion and exercise training on A and B: whole-cell K^+ channel current and C and D: iberiotoxin-sensitive K^+ channel current in smooth muscle cells from coronary arteries and arterioles. Currents were elicited by 500-ms step depolarizations to test potentials (TP) ranging from −70 to +100 in 10-mV increments from a holding potential of −80 mV. Current-voltage relationships reveal relatively smaller whole-cell K^+ channel currents in smooth muscle cells isolated from conduit compared with arterioles in nonoccluded (A) and collateral-dependent (B) vessels. In contrast, iberiotoxin-sensitive K^+ currents are relatively larger in conduit than arteriole smooth muscle cells of both nonoccluded (C) and collateral-dependent (D) vessels. Current-voltage relationships were obtained by plotting the mean outward current at the end of each test potential normalized to cell membrane capacitance. Values are means ± SEM; *n*, number of animals, followed by number of cells. The legend key on left represents panels A and C; legend key on right represents panel B and D. *Data were adapted from Deer, R. R., & Heaps, C. L. (2013). Exercise training enhances multiple mechanisms of relaxation in coronary arteries from ischemic hearts.* American Journal of Physiology. Heart and Circulatory Physiology, 305, *H1321-H1331.; W. Xie, W., Parker, J. L., & Heaps, C. L. (2013). Exercise training-enhanced, endothelium-dependent dilation mediated by altered regulation of BKCa channels in collateral-dependent porcine coronary arterioles.* Microcirculation, 20(2), *170–182.*

a solution containing high EGTA (10 mM), known to chelate intracellular Ca^{2+} and thereby minimize BKCa channel activity to that stimulated by changes in voltage; thus, while currents are comparable, the iberiotoxin-sensitive K^+ currents are relatively low.

Interestingly, despite no exercise training-induced alterations in BKCa channel currents, these channels did contribute significantly to exercise training-enhanced, endothelium-dependent dilation of the coronary microvasculature (Xie et al., 2013). In cannulated coronary arterioles, endothelium-dependent, bradykinin-mediated dilation was impaired in the collateral-dependent region of ameroid-occluded swine (Xie et al., 2013). Exercise training restored the impaired dilation, which we revealed was at least partially dependent on a greater contribution of BKCa channels after exercise (Xie et al., 2013). We also demonstrated in coronary arteries, that although bradykinin-mediated relaxation was not altered by occlusion or exercise training under control conditions, relaxation was more persistent after inhibition of the endothelium-derived signaling molecules, nitric oxide and prostacyclin, after exercise training (Deer & Heaps, 2013). Inclusion of the BKCa channel blocker, iberiotoxin, with inhibitors of nitric oxide synthase and prostanoids abolished the enhanced exercise-training-mediated relaxation in nonoccluded and collateral-dependent arteries, such that relaxation responses were similar across treatments (Deer & Heaps, 2013). These findings suggested redundancy in signaling pathways of vascular relaxation after exercise training and that BKCa channels act in a compensatory role when other signaling molecules are impaired. In addition, our laboratory as well as others have reported involvement of BKCa channels in coronary vascular recovery and enhanced relaxation after exercise training in cardiovascular disease models (Deer & Heaps, 2013; Xie et al., 2013).

Recent data from our laboratory have revealed that exercise training restores impaired endothelium-dependent dilation in collateral-dependent arterioles through cellular adaptations that increase endothelium-derived H_2O_2 bioavailability (Xie, Parker, & Heaps, 2012). Our subsequent data indicate that BKCa and 4-aminopyridine-sensitive Kv channels are downstream effectors that contributed to enhanced H_2O_2-mediated dilation in the collateral-dependent arterioles after exercise training (Johnson, Bray, & Heaps, 2021). Using confocal microscopy, we also demonstrated that H_2O_2 treatment increased colocalization of protein kinase A and BKCa channel proteins at the plasma membrane of smooth muscle cells isolated from collateral-dependent arterioles of exercise trained pigs (Johnson et al., 2021). These findings implicate increased PKA activation of BKCa channels as the signaling pathway that mediates the enhanced H_2O_2-induced dilation after exercise training in arterioles isolated from ischemic myocardium. We postulate that PKA activation of Kv channels may also play a role in the enhanced dilation observed after exercise training. Future studies will explore the contribution of specific Kv channel isoforms as downstream effectors of enhanced H_2O_2-mediated dilation after exercise training.

6. Role of additional K^+ channel subfamilies in control of basal tone and vasodilation

Our laboratory has not explored the effect of chronic occlusion and exercise training on KATP channels in the coronary microcirculation. It has been shown that inhibition of KATP channels leads to increased basal tone, suggesting a significant contribution of these channels to the regulation of resting coronary blood flow (Duncker, Van Zon, Altman, Pavek, & Bache, 1993; Duncker, van Zon, Pavek, Herrlinger, & Bache, 1995; Merkus, Sorop, Houweling, Hoogteijling, & Duncker, 2006; Samaha, Heineman, Ince, Fleming, & Balaban, 1992). In human patients, Farouque, Worthley, Meredith, Skyrme-Jones, and Zhang (2002) assessed coronary hemodynamics before and after KATP blockade with intracoronary delivery of glibenclamide. In this study, glibenclamide significantly reduced resting coronary blood flow by 17%, but did not alter coronary flow reserve. However, while KATP channels play a significant role in determination of basal tone in swine and humans, several studies have demonstrated that they are not necessary for vasodilation associated with exercise hyperemia in animal models (Duncker, Oei, Hu, Stubenitsky, & Verdouw, 2001; Duncker et al., 1993; Duncker et al., 1995; Merkus et al., 2003) or adenosine-induced coronary hyperemia in humans (Farouque et al., 2002). We previously reported no role for KATP channels in adenosine-mediated dilation of coronary arterioles isolated from control male or female swine (Heaps & Bowles, 2002). Other investigators suggest a role for KATP channels after transient ischemia, i.e., reactive hyperemia (Berwick et al., 2010) and under hypoxic conditions (Liu & Flavahan, 1997), but the role of these channels in animal models or human patients with critical coronary stenosis is unclear.

In contrast to BKCa channels, small-conductance (SKCa) and intermediate-conductance (IKCa) are predominantly expressed within endothelial cells and have been shown to demonstrate higher sensitivity to calcium in comparison to BKCa channels (Burnham et al., 2002). Substantial literature support endothelial IKCa and SKCa channels as contributors to endothelium-derived hyperpolarization and vascular tone regulation (Levi-Rosenzvig et al., 2017; Liu et al., 2015). Additional studies have also revealed the expression of IKCa channels in vascular smooth muscle, however, this occurs primarily within proliferating smooth muscle cells (Bi et al., 2013; Tharp, Wamhoff, Turk, & Bowles, 2006; Zhao, Su, Wang, Li, & Deng, 2013), playing a role in the pathogenesis of cardiovascular disease and restenosis. Yet, Gutterman and colleagues (Sato et al., 2005) reported a significant role of smooth muscle IKCa channels in adenosine-mediated dilation with no significant contribution of BKCa, SKCa, KATP, or 4-AP-sensitive Kv channels in coronary arterioles of diseased patients. Within endothelial cells, Tune and colleagues (Kurian, Berwick, & Tune, 2011) found that activation of coronary microvascular IKCa channels resulted in increased coronary blood flow via nitric oxide production, whereas inhibition of the IKCa channels resulted in impaired coronary flow regulation. However, inhibition of these channels did not contribute to dilation following brief occlusion. In a hypoxic environment, other investigators found reduced endothelial cell K^+ currents, mainly involving reduced SKCa and IKCa channels in porcine coronary arteries (Yang, Huang, Man, Yao, & He, 2011). These reductions were accompanied by decreased IKCa channel protein levels. Thus, SKCa and IKCa channels contribute mostly to endothelium-derived vasodilation with a negative impact of hypoxia; adaptations to chronic ischemia have not been explored.

7. Conclusion

This review has primarily focused on adaptations in K^+ channel activity and contributions to vasodilation in the microcirculation distal to chronic coronary artery occlusion. We propose that K^+ channel activity is a highly significant player in the dynamic regulation of coronary blood flow under physiological conditions as well as in adaptations to disease and exercise training. Despite our focus on adaptations in cellular signaling in the microcirculation distal to stenosis, we postulate that some of the observed alterations may be applicable to coronary microvascular dysfunction in the absence of obstructive coronary artery disease. Further research and insight into mechanisms that control coronary blood flow in health and disease, as well as exercise training will support novel and potentially more effective therapeutic approaches for the prevention and treatment of cardiovascular disorders.

References

Ahn, S. J., Fancher, I. S., Bian, J. T., Zhang, C. X., Schwab, S., Gaffin, R., et al. (2017). Inwardly rectifying K(+) channels are major contributors to flow-induced vasodilatation in resistance arteries. *The Journal of Physiology*, *595*(7), 2339–2364. https://doi.org/10.1113/jp273255.

Albarwani, S., Al-Siyabi, S., Baomar, H., & Hassan, M. O. (2010). Exercise training attenuates ageing-induced BKCa channel downregulation in rat coronary arteries. *Experimental Physiology*, *95*(6), 746–755. https://doi.org/10.1113/expphysiol.2009.051250.

Armstrong, R. B., Delp, M. D., Goljan, E. F., & Laughlin, M. H. (1987). Distribution of blood flow in muscles of miniature swine during exercise. *Journal of Applied Physiology*, *62*(3), 1285–1298.

Bache, R. J., & Schwartz, J. S. (1983). Myocardial blood flow during exercise after gradual coronary occlusion in the dog. *American Journal of Physiology. Heart and Circulatory Physiology*, *245*(1), H131–H138.

Belardinelli, R., Belardinelli, L., & Shryock, J. C. (2001). Effects of dipyridamole on coronary collateralization and myocardial perfusion in patients with ischaemic cardiomyopathy. *European Heart Journal*, *22*(14), 1205–1213.

Belardinelli, R., Georgiou, D., Ginzton, L., Cianci, G., & Purcaro, A. (1998). Effects of moderate exercise training on thallium uptake and contractile response to low-dose dobutamine of dysfunctional myocardium in patients with ischemic cardiomyopathy. *Circulation*, *97*(6), 553–561.

Berkefeld, H., Fakler, B., & Schulte, U. (2010). Ca2+—activated K+ channels: From protein complexes to function. *Physiological Reviews*, *90*(4), 1437–1459. https://doi.org/10.1152/physrev.00049.2009.

Berwick, Z. C., Payne, G. A., Lynch, B., Dick, G. M., Sturek, M., & Tune, J. D. (2010). Contribution of adenosine A(2A) and A(2B) receptors to ischemic coronary dilation: Role of K(V) and K(ATP) channels. *Microcirculation*, *17*(8), 600–607. https://doi.org/10.1111/j.1549-8719.2010.00054.x.

Bi, D., Toyama, K., Lemaître, V., Takai, J., Fan, F., Jenkins, D. P., et al. (2013). The intermediate conductance calcium-activated potassium channel KCa3.1 regulates vascular smooth muscle cell proliferation via controlling calcium-dependent signaling. *The Journal of Biological Chemistry*, *288*(22), 15843–15853. https://doi.org/10.1074/jbc.M112.427187.

Bloor, C. M., White, F. C., & Sanders, T. M. (1984). Effects of exercise on collateral development in myocardial ischemia in pigs. *Journal of Applied Physiology*, *56*, 656–665.

Bowles, D. K., Hu, Q., Laughlin, M. H., & Sturek, M. (1998). Exercise training increases L-type calcium current density in coronary smooth muscle. *The American Journal of Physiology*, *275*, H2159–H2169.

Bowles, D. K., Laughlin, M. H., & Sturek, M. (1998). Exercise training increases K^+ channel contribution to regulation of coronary arterial tone. *Journal of Applied Physiology*, *84*(4), 1225–1233.

Burnham, M. P., Bychkov, R., Félétou, M., Richards, G. R., Vanhoutte, P. M., Weston, A. H., et al. (2002). Characterization of an apamin-sensitive small-conductance Ca(2+)-activated K(+) channel in porcine coronary artery endothelium: Relevance to EDHF. *British Journal of Pharmacology*, *135*(5), 1133–1143. https://doi.org/10.1038/sj.bjp.0704551.

Cohen, M. V. (1985). Functional significance of coronary collaterals in man. In *Coronary collaterals: Clinical and experimental observation* (pp. 93–185). New York: Futura Publishing Co. reprinted from: In file.

Cohen, M. V., & Yipintsoi, T. (1981). Restoration of cardiac function and myocardial flow by collateral development in dogs. *American Journal of Physiology. Heart and Circulatory Physiology*, *240*(5), H811–H819.

Cohen, M. V., Yipintsoi, T., & Scheuer, J. (1982). Coronary collateral stimulation by exercise in dogs with stenotic coronary arteries. *Journal of Applied Physiology, 52*(3), 664–671.

Conraads, V. M., Pattyn, N., De Maeyer, C., Beckers, P. J., Coeckelberghs, E., Cornelissen, V. A., et al. (2015). Aerobic interval training and continuous training equally improve aerobic exercise capacity in patients with coronary artery disease: The SAINTEX-CAD study. *International Journal of Cardiology, 179*, 203–210. https://doi.org/10.1016/j.ijcard.2014.10.155.

Dart, C., & Standen, N. B. (1995). Activation of ATP-dependent K+ channels by hypoxia in smooth muscle cells isolated from the pig coronary artery. *The Journal of Physiology, 483*(Pt 1), 29–39. https://doi.org/10.1113/jphysiol.1995.sp020565.

Deer, R. R., & Heaps, C. L. (2013). Exercise training enhances multiple mechanisms of relaxation in coronary arteries from ischemic hearts. *American Journal of Physiology. Heart and Circulatory Physiology, 305*, H1321–H1331.

Douglas, W. R. (1972). Of pigs and men and research. *Space Life Sciences, 3*, 226–234.

Duncker, D. J., & Bache, R. J. (2008). Regulation of coronary blood flow during exercise. *Physiological Reviews, 88*(3), 1009–1086.

Duncker, D. J., Koller, A., Merkus, D., & Canty, J. M., Jr. (2015). Regulation of coronary blood flow in health and ischemic heart disease. *Progress in Cardiovascular Diseases, 57*(5), 409–422. https://doi.org/10.1016/j.pcad.2014.12.002.

Duncker, D. J., Oei, H. H., Hu, F., Stubenitsky, R., & Verdouw, P. D. (2001). Role of K(ATP)(+) channels in regulation of systemic, pulmonary, and coronary vasomotor tone in exercising swine. *American Journal of Physiology. Heart and Circulatory Physiology, 280*(1), H22–H33. https://doi.org/10.1152/ajpheart.2001.280.1.H22.

Duncker, D. J., Van Zon, N. S., Altman, J. D., Pavek, T. J., & Bache, R. J. (1993). Role of K+ATP channels in coronary vasodilation during exercise. *Circulation, 88*(3), 1245–1253. https://doi.org/10.1161/01.cir.88.3.1245.

Duncker, D. J., van Zon, N. S., Pavek, T., Herrlinger, S. K., & Bache, R. J. (1995). Endogenous adenosine mediates coronary vasodilation during exercise after K+ATP channel blockade. *The Journal of Clinical Investigation, 95*, 285–295.

Dwenger, M. M., Ohanyan, V., Navedo, M. F., & Nystoriak, M. A. (2018). Coronary microvascular Kv1 channels as regulatory sensors of intracellular pyridine nucleotide redox potential. *Microcirculation, 25*(1). https://doi.org/10.1111/micc.12426.

Eckstein, R. W. (1957). Effect of exercise and coronary artery narrowing on coronary collateral circulation. *Circulation Research, 5*, 230–235.

Fancher, I. S., Ahn, S. J., Adamos, C., Osborn, C., Oh, M. J., Fang, Y., et al. (2018). Hypercholesterolemia-induced loss of flow-induced vasodilation and lesion formation in apolipoprotein E-deficient mice critically depend on inwardly rectifying K(+) channels. *Journal of the American Heart Association, 7*(5). https://doi.org/10.1161/jaha.117.007430.

Farouque, H. M., Worthley, S. G., Meredith, I. T., Skyrme-Jones, R. A., & Zhang, M. J. (2002). Effect of ATP-sensitive potassium channel inhibition on resting coronary vascular responses in humans. *Circulation Research, 90*(2), 231–236. https://doi.org/10.1161/hh0202.103713.

GBD 2015 Disease and Injury Incidence and Prevalence Collaborators. (2016). Global, regional, and national incidence, prevalence, and years lived with disability for 310 diseases and injuries, 1990–2015: A systematic analysis for the global burden of disease study 2015. *Lancet, 388*(10053), 1545–1602. https://doi.org/10.1016/s0140-6736(16)31678-6.

Gebremedhin, D., Bonnet, P., Greene, A. S., England, S. K., Rusch, N. J., Lombard, J. H., et al. (1994). Hypoxia increases the activity of Ca(2+)-sensitive K+ channels in cat cerebral arterial muscle cell membranes. *Pflügers Archiv, 428*(5–6), 621–630. https://doi.org/10.1007/bf00374586.

Goodwill, A. G., Fu, L., Noblet, J. N., Casalini, E. D., Sassoon, D., Berwick, Z. C., et al. (2016). KV7 channels contribute to paracrine, but not metabolic or ischemic, regulation of coronary vascular reactivity in swine. *American Journal of Physiology. Heart and Circulatory Physiology*, *310*(6), H693–H704. https://doi.org/10.1152/ajpheart.00688.2015.

Griffin, K. L., Laughlin, M. H., & Parker, J. L. (1999). Exercise training improves endothelium-mediated vasorelaxation after chronic coronary occlusion. *Journal of Applied Physiology*, *87*(5), 1948–1956.

Griffin, K. L., Woodman, C. R., Price, E. M., Laughlin, M. H., & Parker, J. L. (2001). Endothelium-mediated relaxation of porcine collateral-dependent arterioles is improved by exercise training. *Circulation*, *104*, 1393–1398.

Gutterman, D. D., Miura, H., & Liu, Y. (2005). Redox modulation of vascular tone: Focus of potassium channel mechanisms of dilation. *Arteriosclerosis, Thrombosis, and Vascular Biology*, *25*(4), 671–678. https://doi.org/10.1161/01.ATV.0000158497.09626.3b.

Hambrecht, R., Wolf, A., Gielen, S., Linke, A., Hofer, J., Erbs, S., et al. (2000). Effect of exercise on coronary endothelial function in patients with coronary artery disease. *The New England Journal of Medicine*, *342*(7), 454–460.

Heaps, C. L., & Bowles, D. K. (2002). Gender-specific K^+-channel contribution to adenosine-induced relaxation in coronary arterioles. *Journal of Applied Physiology*, *92*, 550–558.

Heaps, C. L., Mattox, M. L., Kelly, K. A., Meininger, C. J., & Parker, J. L. (2006). Exercise training increases basal tone in arterioles distal to chronic coronary occlusion. *American Journal of Physiology. Heart and Circulatory Physiology*, *290*, H1128–H1135.

Heaps, C. L., & Parker, J. L. (2011). Effects of exercise training on coronary collateralization and control of collateral resistance. *Journal of Applied Physiology*, *111*(2), 587–598.

Hedegaard, E. R., Nielsen, B. D., Kun, A., Hughes, A. D., Krøigaard, C., Mogensen, S., et al. (2014). KV 7 channels are involved in hypoxia-induced vasodilatation of porcine coronary arteries. *British Journal of Pharmacology*, *171*(1), 69–82. https://doi.org/10.1111/bph.12424.

Hibino, H., Inanobe, A., Furutani, K., Murakami, S., Findlay, I., & Kurachi, Y. (2010). Inwardly rectifying potassium channels: Their structure, function, and physiological roles. *Physiological Reviews*, *90*(1), 291–366. https://doi.org/10.1152/physrev.00021.2009.

Hill, M. A., Yang, Y., Ella, S. R., Davis, M. J., & Braun, A. P. (2010). Large conductance, Ca2+-activated K+ channels (BKCa) and arteriolar myogenic signaling. *FEBS Letters*, *584*(10), 2033–2042. https://doi.org/10.1016/j.febslet.2010.02.045.

Jackson, W. F. (2018). K(V) channels and the regulation of vascular smooth muscle tone. *Microcirculation*, *25*(1), 1–13. https://doi.org/10.1111/micc.12421.

Johnson, K., Bray, J., & Heaps, C. (2021). Exercise training rescues impaired H2O2-mediated vasodilation in collateral-dependent coronary arterioles through enhanced K+ channel activation. *The FASEB Journal*, *35*(S1). https://doi.org/10.1096/fasebj.2021.35.S1.02016.

Kamekura, I., Okumura, K., Matsui, H., Murase, K., Mokuno, S., Toki, Y., et al. (1999). Mechanisms of hypoxic coronary vasodilatation in isolated perfused rat hearts. *Journal of Cardiovascular Pharmacology*, *33*(6), 836–842. https://doi.org/10.1097/00005344-199906000-00002.

Kim, H. R., Appel, S., Vetterkind, S., Gangopadhyay, S. S., & Morgan, K. G. (2008). Smooth muscle signalling pathways in health and disease. *Journal of Cellular and Molecular Medicine*, *12*(6a), 2165–2180. https://doi.org/10.1111/j.1582-4934.2008.00552.x.

Kolibash, A. J., Bush, C. A., Wepsic, R. A., Schroeder, D. P., Tetalman, M. R., & Lewis, R. P. (1982). Coronary collateral vessels: Spectrum of physiologic capabilities with respect to providing rest and stress myocardial perfusion, maintenance of left ventricular function and protection against infarction. *The American Journal of Cardiology*, *50*, 230–238.

Koller, A., Laughlin, M. H., Cenko, E., de Wit, C., Tóth, K., Bugiardini, R., et al. (2022). Functional and structural adaptations of the coronary macro- and microvasculature to regular aerobic exercise by activation of physiological, cellular, and molecular mechanisms: ESC working group on coronary pathophysiology and microcirculation position paper. *Cardiovascular Research, 118*(2), 357–371. https://doi.org/10.1093/cvr/cvab246.

Kurian, M. M., Berwick, Z. C., & Tune, J. D. (2011). Contribution of IKCa channels to the control of coronary blood flow. *Experimental Biology and Medicine (Maywood, N.J.), 236*(5), 621–627. https://doi.org/10.1258/ebm.2011.010351.

Lambert, P. R., Hess, D. S., & Bache, R. J. (1977). Effect of exercise on perfusion of collateral-dependent myocardium in dogs with chronic coronary artery occlusion. *Journal of Clinical Investigation, 59*(1), 1–7.

Laughlin, M. H., Bowles, D. K., & Duncker, D. J. (2012). The coronary circulation in exercise training. *American Journal of Physiology. Heart and Circulatory Physiology, 302*(1), H10–H23. https://doi.org/10.1152/ajpheart.00574.2011.

Lee, Y. H., Kim, J. T., & Kang, B. S. (1998). Mechanisms of relaxation of coronary artery by hypoxia. *Yonsei Medical Journal, 39*(3), 252–260. https://doi.org/10.3349/ymj.1998.39.3.252.

Lee, S. H., Shin, D., Lee, J. M., van de Hoef, T. P., Hong, D., Choi, K. H., et al. (2022). Clinical relevance of ischemia with nonobstructive coronary arteries according to coronary microvascular dysfunction. *Journal of the American Heart Association, 11*(9), e025171. https://doi.org/10.1161/jaha.121.025171.

Levi-Rosenzvig, R., Beyer, A. M., Hockenberry, J., Ben-Shushan, R. S., Chuyun, D., Atiya, S., et al. (2017). 5,6-δ-DHTL, a stable metabolite of arachidonic acid, is a potential EDHF that mediates microvascular dilation. *Free Radical Biology & Medicine, 103*, 87–94. https://doi.org/10.1016/j.freeradbiomed.2016.12.022.

Li, Z., Lu, N., & Shi, L. (2014). Exercise training reverses alterations in Kv and BKCa channel molecular expression in thoracic aorta smooth muscle cells from spontaneously hypertensive rats. *Journal of Vascular Research, 51*(6), 447–457. https://doi.org/10.1159/000369928.

Liu, Y., Bubolz, A. H., Mendoza, S., Zhang, D. X., & Gutterman, D. D. (2011). H_2O_2 is the transferrable factor mediating flow-induced dilation in human coronary arterioles. *Circulation Research, 108*(5), 566–573.

Liu, Q., & Flavahan, N. A. (1997). Hypoxic dilatation of porcine small coronary arteries: Role of endothelium and KATP-channels. *British Journal of Pharmacology, 120*(4), 728–734. https://doi.org/10.1038/sj.bjp.0700939.

Liu, Y., Xie, A., Singh, A. K., Ehsan, A., Choudhary, G., Dudley, S., et al. (2015). Inactivation of endothelial small/intermediate conductance of calcium-activated potassium channels contributes to coronary arteriolar dysfunction in diabetic patients. *Journal of the American Heart Association, 4*(8), e002062. https://doi.org/10.1161/jaha.115.002062.

Longhurst, J. C., Motohara, S., Atkins, J. M., & Ordway, G. A. (1985). Function of mature coronary collateral vessels and cardiac performance in the exercising dog. *Journal of Applied Physiology, 59*(2), 392–400.

López-Barneo, J., del Toro, R., Levitsky, K. L., Chiara, M. D., & Ortega-Sáenz, P. (2004). Regulation of oxygen sensing by ion channels. *Journal of Applied Physiology (Bethesda, MD: 1985), 96*(3), 1187–1195. discussion 1170-1182 https://doi.org/10.1152/japplphysiol.00929.2003.

Merkus, D., Haitsma, D. B., Fung, T.-Y., Assen, Y. J., Verdouw, P. D., & Duncker, D. J. (2003). Coronary blood flow regulation in exercising swine involves parallel rather than redundant vasodilator pathways. *American Journal of Physiology. Heart and Circulatory Physiology, 285*, H424–H433.

Merkus, D., Muller-Delp, J., & Heaps, C. L. (2021). Coronary microvascular adaptations distal to epicardial artery stenosis. *American Journal of Physiology. Heart and Circulatory Physiology*, *320*(6), H2351–h2370. https://doi.org/10.1152/ajpheart.00992.2020.

Merkus, D., Sorop, O., Houweling, B., Hoogteijling, B. A., & Duncker, D. J. (2006). KCa+ channels contribute to exercise-induced coronary vasodilation in swine. *American Journal of Physiology. Heart and Circulatory Physiology*, *291*(5), H2090–H2097.

Miura, H., Toyama, K., Pratt, P. F., & Gutterman, D. D. (2011). Cigarette smoking impairs Na+-K+-ATPase activity in the human coronary microcirculation. *American Journal of Physiology. Heart and Circulatory Physiology*, *300*(1), H109–H117. https://doi.org/10.1152/ajpheart.00237.2010.

Miura, H., Wachtel, R. E., Liu, Y., Loberiza, F. R., Jr., Saito, T., Miura, M., et al. (2001). Flow-induced dilation of human coronary arterioles: Important role of Ca(2+)-activated K(+) channels. *Circulation*, *103*(15), 1992–1998.

Mokelke, E. A., Dietz, N. J., Eckman, D. M., Nelson, M. T., & Sturek, M. (2005). Diabetic dyslipidemia and exercise affect coronary tone and differential regulation of conduit and microvessel K+ current. *American Journal of Physiology. Heart and Circulatory Physiology*, *288*(3), H1233–H1241. https://doi.org/10.1152/ajpheart.00732.2004.

Mokelke, E. A., Hu, Q., Song, M., Toro, L., Reddy, H. K., & Sturek, M. (2003). Altered functional coupling of coronary K^+ channels in diabetic dyslipidemic pigs is prevented by exercise. *Journal of Applied Physiology*, *95*, 1179–1193.

Nelson, M. T., Patlak, J. B., Worley, J. F., & Standen, N. B. (1990). Calcium channels, potassium channels, and voltage dependence of arterial smooth muscle tone. *The American Journal of Physiology*, *259*, C3–C18.

Nelson, M. T., & Quayle, J. M. (1995). Physiological roles and properties of potassium channels in arterial smooth muscle. *American Journal of Physiology. Cell Physiology*, *268*, C799–C822.

Niebauer, J., Hambrecht, R., Marburger, C., Hauer, K., Velich, T., von Hodenberg, E., et al. (1995). Impact of intensive physical exercise and low-fat diet on collateral vessel formation in stable angina pectoris and angiographically confirmed coronary artery disease. *The American Journal of Cardiology*, *76*(11), 771–775.

Nishijima, Y., Cao, S., Chabowski, D. S., Korishettar, A., Ge, A., Zheng, X., et al. (2017). Contribution of Kv1.5 channel to H2O2-induced human arteriolar dilation and its modulation by coronary artery disease. *Circulation Research*, *120*(4), 658–669. https://doi.org/10.1161/circresaha.116.309491.

Nishijima, Y., Korishettar, A., Chabowski, D. S., Cao, S., Zheng, X., Gutterman, D. D., et al. (2018). Shaker-related voltage-gated K(+) channel expression and vasomotor function in human coronary resistance arteries. *Microcirculation*, *25*(1). https://doi.org/10.1111/micc.12431.

Ohanyan, V., Yin, L., Bardakjian, R., Kolz, C., Enrick, M., Hakobyan, T., et al. (2015). Requisite role of Kv1.5 channels in coronary metabolic dilation. *Circulation Research*, *117*(7), 612–621. https://doi.org/10.1161/circresaha.115.306642.

O'Konski, M. S., White, F. C., Longhurst, J. C., Roth, D. M., & Bloor, C. M. (1986). Ameroid constriction of the proximal left circumflex coronary artery in swine. *The American Journal of Cardiovascular Pathology*, *1*(1), 69–77.

Olver, T. D., Edwards, J. C., Ferguson, B. S., Hiemstra, J. A., Thorne, P. K., Hill, M. A., et al. (2018). Chronic interval exercise training prevents BK(Ca) channel-mediated coronary vascular dysfunction in aortic-banded miniswine. *Journal of Applied Physiology (Bethesda, MD: 1985)*, *125*(1), 86–96. https://doi.org/10.1152/japplphysiol.01138.2017.

Quayle, J. M., Nelson, M. T., & Standen, N. B. (1997). ATP-sensitive and inwardly rectifying potassium channels in smooth muscle. *Physiological Reviews*, 77(4), 1165–1232.

Rogers, P. A., Chilian, W. M., Bratz, I. N., Bryan, R. M., & Dick, G. M. (2007). H_2O_2 activates redox- and 4-aminopyridine-sensitive Kv channels in coronary vascular smooth muscle. *American Journal of Physiology. Heart and Circulatory Physiology, 292*(3), H1404–H1411.

Rogers, P. A., Dick, G. M., Knudson, J. D., Focardi, M., Bratz, I. N., Swafford, A. N., et al. (2006). H_2O_2-induced redox-sensitive coronary vasodilation is mediated by 4-aminopyridine-sensitive K^+ channels. *American Journal of Physiology. Heart and Circulatory Physiology, 291*(5), H2473–H2482.

Roth, D. M., Maruoka, Y., Rogers, J., White, F. C., Longhurst, J. C., & Bloor, C. M. (1987). Development of coronary collateral circulation in left circumflex ameroid-occluded swine myocardium. *The American Journal of Physiology, 253*, H1279–H1288.

Roth, D. M., White, F. C., Nichols, M. L., Dobbs, S. L., Longhurst, J. C., & Bloor, C. M. (1990). Effects of long-term exercise on regional myocardial function and coronary collateral development after gradual coronary artery occlusion in pigs. *Circulation, 82*, 1778–1789.

Rusch, N. J. (2009). BK channels in cardiovascular disease: A complex story of channel dysregulation. *American Journal of Physiology. Heart and Circulatory Physiology, 297*(5), H1580–H1582. https://doi.org/10.1152/ajpheart.00852.2009.

Samaha, F. F., Heineman, F. W., Ince, C., Fleming, J., & Balaban, R. S. (1992). ATP-sensitive potassium channel is essential to maintain basal coronary vascular tone in vivo. *The American Journal of Physiology, 262*(5 Pt 1), C1220–C1227. https://doi.org/10.1152/ajpcell.1992.262.5.C1220.

Sato, A., Terata, K., Miura, H., Toyama, K., Loberiza, F. R. J., Hatoum, O. A., et al. (2005). Mechanism of vasodilation to adenosine in coronary arterioles from patients with heart disease. *American Journal of Physiology. Heart and Circulatory Physiology, 288*, H1633–H1640.

Schaper, W. (1971a). Comparative arteriography of the collateral circulation. In D. A. K. Black (Ed.), *The collateral circulation of the heart* (pp. 29–50). North-Holland Publishing Company. reprinted from: In file.

Schaper, W. (1971b). The morphology of collaterals and anastomoses in human, canine and porcine hearts. In D. A. K. Black (Ed.), *The collateral circulation of the heart* (pp. 5–18). North-Holland Publishing Company. reprinted from: In file.

Schuler, G., Hambrecht, R., Schlierf, G., Niebauer, J., Hauer, K., Neumann, J., et al. (1992). Regular physical exercise and low-fat diet. Effects on progression of coronary artery disease. *Circulation, 86*(1), 1–11.

Shimizu, S., Yokoshiki, H., Sperelakis, N., & Paul, R. J. (2000). Role of voltage-dependent and Ca^{2+}-activated K^+ channels on the regulation of isometric force in porcine coronary artery. *Journal of Vascular Research, 37*, 16–25.

Sonkusare, S. K., Dalsgaard, T., Bonev, A. D., & Nelson, M. T. (2016). Inward rectifier potassium (Kir2.1) channels as end-stage boosters of endothelium-dependent vasodilators. *The Journal of Physiology, 594*(12), 3271–3285. https://doi.org/10.1113/jp271652.

Tharp, D. L., Wamhoff, B. R., Turk, J. R., & Bowles, D. K. (2006). Upregulation of intermediate-conductance Ca2+-activated K+ channel (IKCa1) mediates phenotypic modulation of coronary smooth muscle. *American Journal of Physiology. Heart and Circulatory Physiology, 291*(5), H2493–H2503. https://doi.org/10.1152/ajpheart.01254.2005.

Thorne, G. D., Conforti, L., & Paul, R. J. (2002). Hypoxic vasorelaxation inhibition by organ culture correlates with loss of Kv channels but not Ca^{2+} channels. *American Journal of Physiology. Heart and Circulatory Physiology, 283*, H247–H253.

Toro, L., Li, M., Zhang, Z., Singh, H., Wu, Y., & Stefani, E. (2014). MaxiK channel and cell signalling. *Pflügers Archiv, 466*(5), 875–886. https://doi.org/10.1007/s00424-013-1359-0.

Tykocki, N. R., Boerman, E. M., & Jackson, W. F. (2017). Smooth muscle ion channels and regulation of vascular tone in resistance arteries and arterioles. *Comprehensive Physiology*, 7(2), 485–581. https://doi.org/10.1002/cphy.c160011.

Vancheri, F., Longo, G., Vancheri, S., & Henein, M. (2020). Coronary microvascular dysfunction. *Journal of Clinical Medicine*, *9*(9). https://doi.org/10.3390/jcm9092880.

White, F. C., & Bloor, C. M. (1981). Coronary collateral circulation in the pig: Correlation of collateral flow with coronary bed size. *Basic Research in Cardiology*, *76*, 189–196.

White, F. C., & Bloor, C. M. (1992). Coronary vascular remodeling and coronary resistance during chronic ischemia. *The American Journal of Cardiovascular Pathology*, *4*(3), 193–202.

White, F. C., Carroll, S. M., Magnet, A., & Bloor, C. M. (1992). Coronary collateral development in swine after coronary artery occlusion. *Circulation Research*, *71*, 1490–1500.

White, F. C., Roth, D. M., & Bloor, C. M. (1986). The pig as a model for myocardial ischemia and exercise. *Laboratory Animal Science*, *36*(4), 351–356.

Wilson, J. R., Kapoor, S. C., & Krishna, G. G. (1994). Contribution of potassium to exercise-induced vasodilation in humans. *Journal of Applied Physiology (Bethesda, MD: 1985)*, 77(6), 2552–2557. https://doi.org/10.1152/jappl.1994.77.6.2552.

Xie, W., Parker, J. L., & Heaps, C. L. (2012). Effect of exercise training on nitric oxide and superoxide/H_2O_2 signaling pathways in collateral-dependent porcine coronary arterioles. *Journal of Applied Physiology*, *112*, 1546–1555.

Xie, W., Parker, J. L., & Heaps, C. L. (2013). Exercise training-enhanced, endothelium-dependent dilation mediated by altered regulation of BK_{Ca} channels in collateral-dependent porcine coronary arterioles. *Microcirculation*, *20*(2), 170–182.

Yang, Q., Huang, J. H., Man, Y. B., Yao, X. Q., & He, G. W. (2011). Use of intermediate/small conductance calcium-activated potassium-channel activator for endothelial protection. *The Journal of Thoracic and Cardiovascular Surgery*, *141*(2), 501–510. 510.e501 https://doi.org/10.1016/j.jtcvs.2010.04.005.

Zhang, D. X., Borbouse, L., Gebremedhin, D., Mendoza, S. A., Zinkevich, N. S., Li, R., et al. (2012). H_2O_2-induced dilation in human coronary arterioles: Role of protein kinase G dimerization and large-conductance Ca^{2+}-activated K^+ channel activation. *Circulation Research*, *110*(3), 471–480.

Zhao, L. M., Su, X. L., Wang, Y., Li, G. R., & Deng, X. L. (2013). KCa3.1 channels mediate the increase of cell migration and proliferation by advanced glycation endproducts in cultured rat vascular smooth muscle cells. *Laboratory Investigation*, *93*(2), 159–167. https://doi.org/10.1038/labinvest.2012.163.